In The Name of God

CONCRETE WORKABILITY

AN INVESTIGATION ON TEMPERATURE EFFECTS USING ARTIFICIAL NEURAL NETWORKS

BY

MOHAMADREZA MOINI

AMIR LAKIZADEH

AuthorHouse™ UK Ltd.
500 Avebury Boulevard
Central Milton Keynes, MK9 2BE
www.authorhouse.co.uk
Phone: 08001974150

First published by AuthorHouse   02/24/2011

ISBN: 978-1-4520-3609-0

Library of Congress Number: 2011901319

ABSTRACT

This book is trying to help expand the knowledge of fresh concrete workability. Concrete workability depends on lots of parameters and varies significantly with conditional alteration. The temperature is of great importance determining the workability. This book presents theoretical information and experimental discussions investigating the influence of fresh concrete temperature and ingredients on the workability. Programming and different statistical programs are applied to include the most appropriate results. Artificial neural network prediction of workability is investigated in two situations; with and without concrete temperature. The most relevant cutting edge studies on this issue are cited and discussed to support the readers for the future studies.

SOMMAIRE

Ce livre essaie de faire progresser la connaissance de la fluidité du béton frais. La fluidité du béton dépend de beaucoup de paramètres et varie de manière significative avec le changement des conditions (température, humidité, souffle du vent, lumière du soleil). La température a beaucoup d'importances pour déterminer la fluidité. Ce livre présente des informations théoriques et des discussions expérimentales pour examiner l'influence de la température du béton frais et les ingrédients sur la fluidité. La programmation et les différents programmes statistiques ont été appliqués afin d'obtenir les résultats les plus appropriés. La prédiction de fluidité de RNA (le réseau de neurones artificiels) a été étudié dans deux situations: avec et sans température du béton frais. Les plus pertinentes recherches, concernant ce sujet, ont été faites pour appuyer les lecteurs aux études futures.

ACKNOWLEDGMENTS

Most things in this world are accomplished by groups rather than by single individuals working alone. The authors express their sincere gratitude to the many people and organizations who have contributed to this book. Special Thanks to Ruhollah Dirbaz our good friend, who had very helpful cooperation and provided constructive criticism and assistance as the head of the concrete laboratory of Qom University. We would also like to thank a number of people who provided us with intellectual support. Majid Shekarian, my mentor and my dear sister who were involved in editing this book. Pazhoohesh Co., Ltd also deserves tremendous thanks for its support through the publication process. Lastly, I would like to appreciate my mother for everything. I remain forever in her debt.

DEDICATED TO

AHMAD & **EHSAN** WHO JOINED
ETERNAL LIFE

Table of Contents

CHAPTER FIVE...........................73

ANALYSIS AND DISCUSSION

CHAPTER SIX...........................91

COCLUSION AND RECOMMENDATIONS

List of Figures

List of Tables

CHAPTER ONE

INTRODUCTION AND BACKGROUND

1.1   INTRODUCTION

The manufacture of concrete with Portland cement in adverse weather conditions involving high and low temperatures directly influences the workability and performance of the fresh concrete during mixing, transport, casting and curing, and its physical and mechanical properties. This is of concern to both concrete manufacturers and final users, because of its affect on a range of technical and economic aspects [1].

A high ambient temperature causes a higher water demand of the concrete and increases the temperature of the fresh concrete. This, results in an increased rate of slump loss and in a more rapid hydration, which leads to accelerated setting and also to a lower long-term strength of concrete [2]. Besides, an increased rate of evaporation from fresh concrete results in a lower effective water content and hence lowers concrete workability. This implies either an addition of water in order to restore the workability or an insufficient compacting or finishing [3]. Likewise in hot climates, there is a tendency for plastic cracking and crazing. As a result, a high temperature can also adversely affect the mechanical properties and serviceability of hardened concrete [4]. Such important effects of temperature on fresh and hardened concrete has made the temperature an influential factor in modeling workability behavior and have triggered numerous studies on various concretes [5,6,7]. Beside its direct effect on workability, the temperature variation of fresh concrete usually result in changes of other ingredients effect on the slump flow.

Artificial neural networks (ANN) are computing systems that simulate the biological neural systems of human brain. They are based on a simplified modeling of the brain's biological functions exhibiting the ability to learn, think, remember, reason, and solve problems. Neural networks have strong and effective abilities to efficiently solve problems that are either difficult or impossible to solve using 'conventional' techniques [15].

In recent years the artificial neural network (ANN) has commonly applied in modeling material behavior [8,9,10] especially for properties of high performance concrete. Neural networks (NNs) have strong and effective abilities to efficiently solve problems that are either difficult or impossible to solve using 'conventional' techniques [11]. NNs are ideally suited for data oriented problems because they can be trained to find latent relationships in data [12]. Due to ability of prediction and therefore reduction of required experiments by learning complex cause and effect relationships from historical experimental data, it facilitates economy.

1.2   PROBLEM STATEMENT

Workability of concrete shares with compressive strength the distinction of being related to diverse types of factors. Each concrete ingredient has an effect on workability. Ingredients proportions interact together to make a distinguished workability. There are much more variants each of which considerably change the workability of concrete. Properties of the cement, properties and proportion of aggregates, amount of air entrained and admixtures and etc. are some examples. Environmental factors such as relative humidity, wind speed, solar radiation, can produce climatic conditions that adversely affect concrete workability. Beside all these, there is one parameter that the concrete workability is also related to. The 'as placed' mixture temperature is a factor which can be assumed as an indication of ambient temperature during placing. There has been little concentration on the issue that to what extent the temperature is influential in predicting workability with high precision using artificial neural networks; Also, how other factors_ basically ingredients_ interact with temperature to change the workability. Determining the accurate and precise correlation between these factors is not possible except using a computer model which can analyze the experimental data. The scope of the present study is primarily focused on the concrete temperature interaction with workability and the amount of mixture temperature influence on predicting workability using ANN models. The following issues are also examined:

1) The interaction between ingredients and workability
2) The interaction between size of coarse aggregates and workability
3) The influence of other parameters on workability

1.3   OBJECTIVES

The main objectives of this research are formulated as follows:

1) To develop an analytical model for concrete workability using ANN;
2) To develop an statistical model for concrete workability using regression analysis;
3) To investigate workability models to find the impact of each mixture variable on workability;
4) To compare behavior of workability models with and without including temperature;
5) To investigate performance of ANN with and without including temperature variable;

This study presents the first stage of experimental-analytical research on concrete workability including mixture temperature interactions with the other factors. The ultimate objective of the research was to investigate to temperature impact on workability, besides simple and practical models to predict workability for a variety of concrete mixtures, including the temperature factors are produced.

1.4   OUTLINE

The present study consists of two major sections; experimentations in which the procedure of providing experiments, mix designs and performing tests are described, and analysis in which the concrete workability models of experimental data are examined and investigated. Both analytical and experimental techniques were employed in this research. This book consists of six chapters beside this introductory chapter. A brief overview of the contents of each chapter follows.

Chapter 2 presents an up-to-date survey of investigations on workability and basics of neural networks. Chapter 3 reviews the method used for experimental and analytical NN section of the research. The NN and statistical methods used for predicting workability is presented in this chapter. This issue is discussed in detail in Chapter 4. It presents the overall results of ANN and statistical programs prediction of workability. The precisions are compared in this chapter. Chapter 5 discusses the detail influence of temperature and ingredients on the workability. Chapter 6 provides conclusions and recommendations for future studies.

CHAPTER TWO

LITERATURE REVIEW

2.1 CONCRETE WORKABILITY

In the construction field, terms like workability, flowability, and cohesion are used sometimes interchangeably to describe the behavior of concrete under flow. American Concrete Institute Standard 116R-90 (ACI 1990b) defines workability as "that property of freshly mixed concrete or mortar which determines the ease and homogeneity, with which it can be mixed, placed, compacted, and finished". Tattersall's [13] interpretation of workability is "the ability of concrete to flow in a mold or formwork, perhaps through congested reinforcement, the ability to be compacted to a minimum volume, perhaps the ability to perform satisfactorily in some transporting operation or forming process, and maybe other requirements as well". Tattersall classifies workability into three classes: qualitative, quantitative empirical and quantitative fundamental as follow [13].

Table 1 Tattersall classification of workability

Class I: qualitative	Class II: quantitative empirical	Class III: quantitative fundamental
Workability, flowability, compactibility, stability, finishability, pumpability, consistency, etc. To be used only in a general descriptive way without any attempt to quantify	To be used as a simple quantitative statement of behavior in a particular set of circumstances	Viscosity, yield stress, etc. To be used in conformity with the British Standard Glossary [14]

Kosmatka et al. [15] mention the following three terms while referring to concrete rheology: workability, consistency and plasticity. The definitions given are:

Table 2 Kosmatka referents of concrete rheology

Workability	Consistency	Plasticity
A measure of how easy or difficult it is to place, consolidate, and finish concrete	The ability of freshly mixed concrete to flow	determines concrete's ease of molding

There are much more similar classifications for workability. It is defined differently by researchers and also is measured by different instruments during decades. These differences have been caused the results of closely different studies cannot be comparative. Although measurements deficiencies are not in the scope of this study, advances in the science of concrete rheology have made fresh concrete workability studies more precise but farther apart because of different instruments and test methods used in measurements.

In common practice, an assumption is made that the standard test for slump of concrete (ASTM C 143) indicates workability. In fact, it correlates well with one component of workability: the yield stress of the concrete. Plastic viscosity also is an essential component of concrete workability but it cannot be indicated by the slump test.

2.2   INFORMATTION SEARCH APPROACH

The principal tool used in the literature review was the search of electronic databases. Keywords used in the search were "concrete," "workability," "temperature," and "neural network." The complete list of reference sources is provided in the reference and bibliography sections of this report.

2.3   FACTORS INFLUENCING WORKABILITY

US Federal Highway Administration has published an extensive report on the concrete workability [16]. Workability is affected by every component of concrete and essentially every condition under which concrete is made. A list of factors includes the properties and the amount of the cement; grading, shape, angularity and surface texture of fine and coarse aggregates; proportion of aggregates; amount of air entrained; type and amount of pozzolan; types and amounts of chemical admixtures; mixing time and method; time since water and cement made contact, and temperature of the concrete mixture [16]. These factors or their combination interact so that changing the proportion of one component to produce a specific characteristic requires that other factors be adjusted to maintain workability. These interactions are discussed extensively in texts and reviews on the subject [17,18,19]. Individual factors are discussed in the following text. In most mixture-proportioning procedures, the water content is assumed to be a factor directly related to the consistency of the concrete for a given maximum size of coarse aggregate. If the water content and the content of cementitious materials are fixed, workability is largely governed by the maximum coarse aggregate size, aggregate shape angularity, texture, and grading. In this study the maximum coarse aggregate size is also included in workability investigations. The coarse-aggregate grading that produces the most workable concrete for one water-cement ratio (w/c) may not produce the most workable concrete for another w/c. As a general rule, the higher the w/c, the finer the aggregate grading is required to produce appropriate flow without segregation.

Three factors in concrete are involved in determining the consistency of the concrete: water-cement ratio, aggregate-cement ratio, and water content. Only two of the three factors are independent. If the aggregate-cement ratio is reduced, the water content must increase for the w/c to remain constant. The water required to maintain a constant consistency will increase as the w/c is increased or decreased [16].

Generally as the fine aggregate/coarse aggregate ratio increase, more water content is required to produce a given workability. If finer aggregate is substituted in a mixture, the water content typically must be increased to maintain the same workability [20]. Similarly, water content must be increased to maintain workability if angular aggregate is substituted for rounded aggregate. Crushed aggregates having numerous flat or elongated particles will produce less workable concrete that requires higher mortar content and possibly a higher paste content. Aggregates with high absorption present a special case because if they are batched with a large unsatisfied absorption, they can remove water from the final concrete mixture and, hence, reduce workability [16].

The size and shape of particles in the fine aggregate affect the workability. For example, the use of very fine sand requires that more water be added to achieve the workability that a coarser sand would provide. Angular fine aggregate particles interlock and reduce the freedom of movement of particles in the fresh concrete. Using angular fine aggregate (e.g., manufactured sand) increases the amount of fine aggregate that must be used for a given amount of coarse aggregate and generally requires that more water be added to achieve the workability obtained with a rounded sand [17].

Lowering the cement content of concrete with a given water content typically will lower workability. A high proportion of cement will produce excellent cohesiveness but may be too sticky to be finished conveniently. An increase in cement fineness decreases workability and produces excessive bleeding, especially when the surface area (Blaine) is less than $280 m^2/kg$. A cement with a high fineness will cause a concrete mixture to lose workability more rapidly because of its rapid hydration [16].

The workability of concrete mixtures commonly is improved by using air-entraining and water-reducing admixtures [21,22]. Air entrainment typically increases paste volume and improves the consistency of the concrete while reducing bleeding and segregation. Water-reducing admixtures disperse cement particles and improve workability, increasing the consistency and reducing segregation [17]. Small changes in the amounts of chemical admixtures used in a concrete can profoundly affect workability. Some chemical admixtures interact in adverse ways with some portland cements, resulting in accelerated hydration of the portland cement.

Mineral admixtures or pozzolans are used to improve strength, durability, and workability in concrete [23]. Freshly mixed concretes are generally more workable when a portion of the cementitious material is fly ash, in part because of the spherical shape of fly ash particles. Smoother mixtures are typically produced if the mineral admixture is substituted for sand rather than cement, but highly reactive or cementitious pozzolans can cause loss of workability through early hydration [17]. Very finely divided mineral admixtures, such as silica fume, can have a very strong negative effect on water demand and hence workability, unless high-range water-reducing admixtures are used [24,25].

Freshly mixed concrete loses workability over time. The reduction in workability is generally attributed to loss of water absorbed into aggregate or by evaporation, or from chemical reaction with the cementitious materials in early hydration reactions [16].

Elevated temperatures increase the rate of water loss in all of the modes mentioned above. The workability of air-entrained concretes is reported to be more easily reduced by elevated temperature than workability in similarly proportioned nonair-entrained concretes [16].

2.4   EVALUATING INFLUENTIAL FACTORS

A list of significant factors including every component of concrete and essentially every producing and placing condition was explained in the previous section. In this section influential factors including temperature, applied to the experiments are independently discussed. Before establishing the correlation between temperature and slump flow in the neural network prediction, a basic description of the effect of each concrete component, the concrete temperature, the MSCA and their interactions on slump flow would be necessary.

2.4.1   Water Content

Water content, independent of other factors is directly proportional to the slump flow at a given temperature. Increasing temperature will lead to increase the rate of water loss and therefore increase of water demand to maintain a constant slump flow. This is cause by increased evaporation, increased viscosity of the liquid-like concrete mixture. Thus the slump changes as the water evaporates from the mixture.

2.4.2   Cement Content

At constant water content the cement content growth will lower the slump. High cement contents, will produce a dry concrete without slump instead of a consistent concrete. The rate of cement hydration is dependent on concrete temperature, admixtures used, etc [4]. The interaction between these factors will specify the slump flow.

2.4.3   Fine Aggregates

Concretes containing high fine aggregates volume fraction of total aggregates reach higher slumps and are more smoothly slumped. The increase in fineness modulus of sand or sand/coarse aggregates ratio generally demands more water content required to produce a given slump. Consequently the rate of slump loss increases at elevated temperatures.

2.4.4   Coarse Aggregates

It is expected that increment of coarse aggregate/ total aggregates ratio decreases workability. Also, the particles size, shape, angularity, surface texture and grading of coarse aggregate are important to determine the water requirement for a given consistency and therefore the slump flow. In this study only amount and maximum size of coarse aggregates are observed in the experiments.

2.4.5   Superplasticizer

Water reducing admixtures decrease water requirement for a consistent concrete and contribute to produce a high performance concrete. The slump increment for the reason of superplasticizer increase, are not similar in varied concrete temperatures.

2.4.6   'As Placed' Temperature

The temperature of concrete in manufacturing process may rise up to even 40°C or below 5°C in some cases. Elevated temperatures increase the rate of water loss and will cause lower slump flow. Increase in temperature during production can transform a workable concrete to a harsh one with poor fluidity. This correlation differs for dissimilar mix portions, different MSCA and different climatic conditions. The temperature also has a strong effect on viscosity of the mixture and friction of aggregates [1]. Different rheological models for fresh concrete behavior have presented numerous relations which assume viscosity either constant or variable [30].

2.4.7   Maximum Size of Coarse Aggregates

According to ACI 211-91 Required mixing water decreases as the maximum size of well-graded aggregate is increased. Concretes with larger-sized aggregates require less mortar per unit volume of concrete. When the MSCA increases, the flow of the cementitious fraction inside the aggregate fraction will be harder and slower. Thus the slump flow decreases with increasing of MSCA when the other factors are held constant.

2.5   WORKABILITY MEASUREMENTS

There are numerous test methods for workability measurements based on rheological understanding of fresh concrete flow. For this study the workability is measured by slump cone test C 143 explanation.

2.6   ARTIFICIAL NEURAL NETWORKS

Artificial neural networks are based on brain's biological functions and attempt to emulate learning processes in the brain though much of the biological detail is neglected. They can learn complex input-output mappings. The learning process is used to determine proper interconnection weights, and the network is trained to properly associate the inputs with their corresponding outputs. The basic strategy for developing a neural-based model of material behavior is to train a neural network on the results of a series of experiments on a material. If the experimental results contain the relevant information about the material behavior, then the trained neural network would contain sufficient information about the material behavior to accept as a material model. Neural networks contain no preconceptions of what the model shape will be, so they are ideal for cases with low system knowledge. They are useful for functional prediction and system modeling where the physical processes are not understood or are highly complex. The generalization capability of the neural networks allows us to predict and produce outputs for the patterns that has been observed before.

Neural networks are composed of simple elements operating in parallel. These elements are inspired by biological nervous systems. As in nature, the connections between elements largely determine the network function.

You can train a neural network to perform a particular function by adjusting the values of the connections (weights) between elements.

Typically, neural networks are adjusted, or trained, so that a particular input leads to a specific target output. Figure 1 illustrates such a situation. There, the network is adjusted, based on a comparison of the output and the target, until the network output matches the target. Typically, many such input/target pairs are needed to train a network.

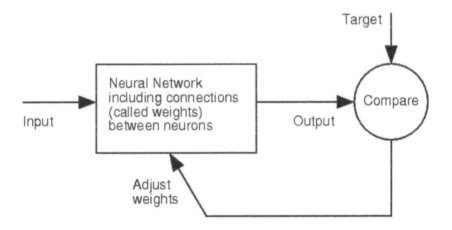

Figure. 1. Adjusting weights in the neural networks

Neural networks can also be trained to solve problems that are difficult for conventional computers or human beings.

2.6.1 Architecture

This section presents the architecture of the networks that are most commonly used with the backpropagation algorithm -- the multilayer feed-forward network.

2.6.1.1   Neuron Model

A network can have several layers. Each layer has a weight matrix W, a bias vector b, and an output vector a. An elementary neuron with R inputs is shown in Figure 2. Each input is weighted with an appropriate w. The sum of the weighted inputs and the bias forms the input to the transfer function f. Neurons can use any differentiable transfer function f to generate their output.

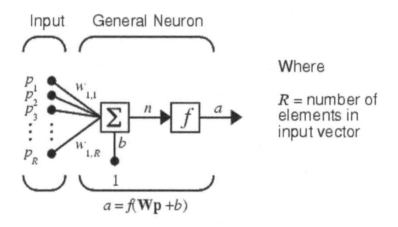

$$a = f(\mathbf{Wp} + b)$$

Figure .2. Inputs and the transfer function of neuron

Multilayer networks often use the log-sigmoid transfer function 'logsig'. The function 'logsig' generates outputs between 0 and 1 as the neuron's net input goes from negative to positive infinity. If the last layer of a multilayer network has sigmoid neurons, then the outputs of the network are limited to a small range. If linear output neurons are used the network outputs can take on any value. 'Logsig', 'tansig', and 'purelin' are the most commonly used transfer functions for backpropagation. Other differentiable transfer functions can be created and used with backpropagation if desired.

2.6.1.2   Feed-forward Network

A single-layer network of S 'logsig' neurons having R inputs is shown in Figure 3 in full detail on the left and with a layer diagram on the right. While two-layer feed-forward networks can potentially learn virtually any input-output relationship, feed-forward networks with more layers might learn complex relationships more quickly.

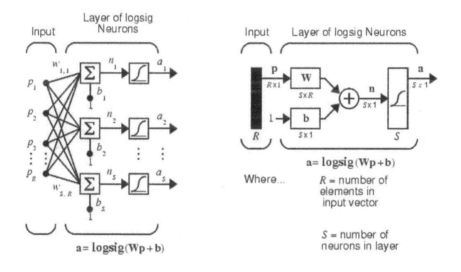

Figure. 3. Typical single layer network

Feed-forward networks often have one or more hidden layers of sigmoid neurons followed by an output layer of linear neurons. Multiple layers of neurons with nonlinear transfer functions allow the network to learn nonlinear and linear relationships between the input and the output vectors. The linear output layer lets the network produce values outside the range -1 to +1.

2.6.1.3   Creating a Network

The first step in training a feed-forward network is to create the network object. It requires three arguments and returns the network object. The first argument is a matrix of sample R-element input vectors. The second argument is a matrix of sample S-element target vectors. The sample inputs and outputs are used to set up network input and output dimensions and parameters. The third argument is an array containing the sizes of each hidden layer. The output layer size is determined from the targets. Before training a feed-forward network, you must initialize the weights and biases. Network inputs might have associated processing functions. Processing functions transform user input data to a form that is easier or more efficient for a network. Some of these functions transform input data so that all values fall into the interval [-1, 1]. This can speed up learning for many networks. Similarly, network outputs can also have associated processing functions. Output processing functions are used to transform user-provided target vectors for network use. Then, network outputs are reverse-processed using the same functions to produce output data with the same characteristics as the original user-provided targets.

2.6.2   Backpropagation

Of the many ANN paradigms, the multi-layer backpropagation network (MLP) is one of the most popular learning algorithms. Backpropagation is the generalization of the Widrow-Hoff learning rule to multiple-layer networks and nonlinear differentiable transfer functions. Input vectors and the corresponding target vectors are used to train a network until it can approximate a function, associate input vectors with specific output vectors, or classify input vectors in an appropriate way.

Standard backpropagation is a gradient descent algorithm, the same as the Widrow-Hoff learning rule, in which the network weights are moved along the negative of the gradient of the performance function. The term backpropagation refers to the manner in which the gradient is computed for nonlinear multilayer networks. There are a number of variations on the basic algorithm that are based on other standard optimization techniques, such as conjugate gradient and Newton methods.

Properly trained backpropagation networks tend to give reasonable answers when presented with inputs that they have never seen. Typically, a new input leads to an output similar to the correct output for input vectors used in training that are similar to the new input being presented. This generalization property makes it possible to train a network on a representative set of input/target pairs and get good results without training the network on all possible input/output pairs.

There are generally four steps in the training process of networks using backpropagation training functions to train feed-forward neural networks for solving specific problems:

1. Assemble the training data.

2. Create the network object.

3. Train the network.

4. Simulate the network response to new inputs.

2.6.2.1   Backpropagation Algorithm

There are many variations of the backpropagation algorithms. The simplest implementation of backpropagation learning updates the network weights and biases in the direction in which the performance function decreases most rapidly, the negative of the gradient. There are two different ways in which this gradient descent algorithm can be implemented: incremental mode and batch mode. In incremental mode, the gradient is computed and the weights are updated after each input is applied to the network. In batch mode, all the inputs are applied to the network before the weights are updated. The batch steepest descent training function is 'traingd'. The weights and biases are updated in the direction of the negative gradient of the performance function. In addition to 'traingd', there are other variations of gradient descent. Gradient descent with momentum, implemented by 'traingdm', allows a network to respond not only to the local gradient, but also to recent trends in the error surface. Gradient descent and gradient descent with momentum are often too slow for practical problems. There are several high-performance algorithms that can converge from ten to one hundred times faster than the mentioned algorithms.

These faster algorithms fall into two categories. The first category uses heuristic techniques, which were developed from an analysis of the performance of the standard steepest descent algorithm. One heuristic modification is the momentum technique, which was presented above. There are two more heuristic techniques: variable learning rate backpropagation, i.e. 'traingda', and resilient backpropagation, i.e. 'trainrp'. The second category of fast algorithms uses standard numerical optimization techniques. Three types of numerical optimization techniques are Conjugate gradient, Quasi-Newton, and Levenberg-Marquardt.

CHAPTER THREE

METHODS AND MATERIALS

3.1   RESEARCH SCOPE AND OBJECTIVES

The primary objective of this investigation is to study the influence of inclusion of concrete temperature in ANN models for predicting the workability determined by slump test. The secondary objective is to evaluate the interaction between temperature and other factors, and the correlation of ingredients with slump when the temperature is applied in the network.

This investigation included a laboratory and an analytical phase. The laboratory phase of this study investigated slump properties of sixty-one conventional concrete mixtures with their temperatures and different maximum size of coarse aggregates (MSCA). The analytical phase of this research used the previously described test method to assess the workability properties. In spite of the limitations inherent in trying to characterize workability by measuring slump, this test method correlates well with one component of workability: the yield stress of the concrete [16]. In this test the slump can be measured by both deduction of the drop from the top of the upright slump cone or by measuring the diameter of the slumped fresh concrete. For the laboratory phase, workability is considered to be measured by diameter of slumped fresh concrete. The following sections present the methods and the materials used in this research program.

3.2   CONCRETE MIXTURES

The concrete mixtures used in the laboratory portion of this research represent a broad range of mixtures. The three maximum size of coarse aggregates (MSCA) used in the laboratory study were 1, ¾ and ½ inch. The fine and coarse aggregates properties for the laboratory mixtures are presented in Appendix A. Properties of the chemical admixture is presented in Appendix B.

3.3   LABORATORY EXPERIMENTAL DESIGN

3.3.1   Tests

The tests developed by using different type of mixtures. Concrete mixtures used in the tests consisted cement, water, coarse aggregates, fine aggregates and superplasticizer. Coarse aggregates also have different maximum size shown in table.3. Amount of superplasticizer in some cases are chosen zero. The temperature of concrete also changes form 4°C to 41°C. These variations in parameters of each mixture help the predicting models to be more general so that the results cover broader range of mixture.

Table 3 A. Range of components of data sets

Component Content	Specific weight	Minimum weight (kg/m^3)	Maximum weight (kg/m^3)
Coarse aggregate	2.64	600	1200
Fine aggregate	2.65	600	1100
Cement	3.15	250	450
Water	1.00	90	247.5
Superplasticizer	1.20	0	20

B. Range of other influential factors

Additional inputs	Minimum amount	Maximum amount	Median
Mixture temperature	4°C	41°C	23°C
MSCA	½ inch	1 inch	¾ inch

C. Range of tested slump value

Tested Value	Minimum amount	Maximum amount	Median
Slump (Diameter)	20 cm	60 cm	34 cm

3.3.2 Testing

A difficult task with the experiments involves the materials that should have been warmed up in different temperatures so that mixing together and produce planned temperature when placing in the slump cone. Achieving desired temperature needed precise calculations and choosing a temperature for each ingredient to be warmed up. Only water, fine and coarse aggregates are heated in the oven to the desired degree of centigrade except for high temperatures, that the cement scarcely warmed up. Besides, to divide coarse aggregate size greater than 1, ¾ and ½ inch such larger particles must have been removed by sieving for the experiments. All the tests were conducted by the author in Qom University Concrete Laboratory. Diameter of slumped flow is measured and the temperature is recorded for each test. All the tests were performed in accordance with ASTM C 192-02.

3.3.3   Test Method

Fresh concrete properties were assessed using ASTM C 143-03, standard test method for slump of hydraulic-cement concrete. A brief summary of the testing procedures follows.

ASTM C 143-03

The mold apparatus of the slump test is filled with concrete in three layers of equal volume. Each layer is compacted with 25 strokes of a tamping rod. The slump cone mold is lifted vertically upward and the change in height of the concrete is measured. The slump test is not considered applicable for concretes with a maximum coarse aggregate size greater than 1.5 inches. For concrete with aggregate greater than 1.5 inches in size, such larger particles can be removed by wet sieving.

3.4   ANALYTICAL INVESTIGATION

Knowledge of predicting concrete properties using artificial neural networks doesn't go far beyond decades [8-12, 26-29]. In many cases in the concrete construction industry slump flow prediction of high performance or conventional concretes, may be required in order to achieve a significant workability. Recent researches have applied slump flow test for ANNs modeling of the workability [11,12]. Nehdi et al. [28] demonstrated that ANN methods can accurately predict the slump flow, filling capacity, and segregation test results of self-compacting concrete. Bai et al. [12] developed ANN models that provide effective predictive capability in respect of the workability of concrete. Yeh [29] demonstrated the abilities of artificial neural networks to represent the effects of each material component on concrete slump. The results show that the models are reliable and accurate. There are much more researches on ANN models predicting strength and durability of HPC or high strength concrete (HSC).

Investigating accuracy of NN slump flow prediction of conventional concretes by including temperature effects is the first time being presented in this study. The input variables for the neural network describe the concrete materials, e.g. cement, water, coarse aggregates, fine aggregates, superplasticizer, the concrete temperature as it is placed in the slump cone and the maximum size of coarse aggregates (MSCA). In this study, the slump flow is a function of these parameters.

To develop the network, good training and testing examples must be obtained experimentally. Table. 3 shows the details of the properties and ranges of ingredients applied for the mixtures. Table. 4 shows the range of ratio of ingredients of the mixtures. The minimum slump flow is the bottom diameter of the slump cone. The database of sixty-one records, used in ANN models are arranged in a format of seven input vectors, and one output value which is a slump flow ranging from 20 to 61.

It must be mentioned that most studies on workability behavior models using ANN are based on data sets of a particular region or country which may produce comparatively different results for a different data set.

Table 4 Range of Ratio of data sets

Ratio	Minimum	Maximum
Fine/Coarse Agg	0.50	1.83
Fine Agg/Cement	2.00	3.80
Coarse/Cement	1.33	4.36
Water/Cement	0.30	0.6
SP/Cement	0.00	0.08
Water/Solid	0.04	0.12
Agg/Cement	3.78	6.80

3.4.1 Neural Network Method for Modeling Slump Flow

A difficult task with ANN involves choosing parameters such as the number of hidden nodes, the learning rate, and the initial weights. Generalization capacity and computational complexity of the network can be directly affected by its topology. Therefore determining appropriate architecture of the network is an important issue.

The neural network used in this study belongs to the class of feed-forward neural networks. The Neural network used here is a five-layer feed-forward, with 7 input neurons corresponding ingredients, 25 neurons in the first hidden layer, 10 neurons in the second hidden layer, and one neuron in the output layer corresponding slump flow of mixtures as shown in Figure 4. After using several topologies, each consist of distinguished hidden layers and learning algorithms, the best accuracy is obtained with this topology utilizing simple trial-and-error method. Training of neural networks was developed under MATLAB programming version 7.6.0. The learning algorithm used in the study was gradient descent with back-propagation, a network training function that updates weight and bias values according to gradient descent. Resilient backpropagation (Rprop) training algorithm was used here for adjusting the weights and 'tansig' function is used as activation function in both hidden layers.

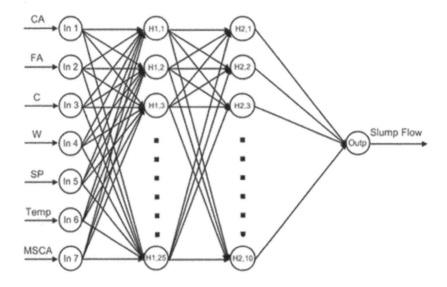

Figure. 4. The architecture of the system based on BPN

In this study 10 fold cross validation method was applied to training and testing models on all the patterns. In this method all 61 patterns were split in 10 sub sets in such a way that all the records were simply shuffled using a random sampling, each sub set contains 6 patterns except the last sub set which contains 7 patterns. Of 10 sub sets, nine sub sets were used for training and one sub set for testing. This process is consequently repeated 10 times, and each time the accuracy for correctly predicted slump flow is recorded. In this way there would be no overlapping between the training and the testing sets. Therefore the neural network is trained and tested on all the patterns of the Database. The final accuracy of 10 tests is the average of accuracies of each test.

The stopping rule of the network is set in a way that for each NN model, weights are iteratively optimized to maximize the number of correctly predicted slump value. In each training process, two small randomly selected categories of the data are set aside for independent test, validation and error estimation in order to avoid possible over fitting. In the end of each epoch, minimum square error on training, testing and validation sets is computed. Iterations for learning, stops if there are more than 10 epochs on validation set in which minimum square error is not decreased.

To avoid floating-point overflow problems the function 'mapstd' is applied to normalize the inputs and targets so that they will have zero mean and unity standard deviation.

To verify the result of NN model, linear and non-linear regression models are also applied to the data set. The next chapter discusses the results of ANN model and compares them with regression analysis results with and without including temperature.

CHAPTER FOUR

PRESENTATION OF
RESULTS

4.1   INTRODUCTION

This section presents the results from the experimental research plan. The results are presented as laboratory and analytical results. Generally two types of laboratory mixtures are produced referring to as Conventional Concrete (CC) and High-performance Concrete (HPC) mixtures. The HPCC mixtures are identified by using supplementary chemical admixture.

4.2   LABORATORY INVESTIGATION

The slump flow of sixty-one laboratory mixtures with different temperatures were assessed using slump cone test and ASTM C 143-03. Temperature measurements were taken at the earliest time for all mixtures. Concrete mixture proportions are presented in Appendix C. Information on the tests and condition of experiments were presented in Section 3, Methods and Materials.

4.2.1   Slump Flow

Slump flow results for laboratory concrete mixtures are presented in Appendix C. The results differ according to amount of superplasticizer used, water, cement, maximum size of coarse aggregates and amount of aggregates. At the highest temperature (41°C) the slump flow diameter is 36 cm. The highest workability is obtained when the mixture temperature is 10°C.

4.3   ANALYTICAL INVESTIGATION

4.3.1   NN Stopping Rule

In a learning process it is anticipated that mean square error (*MSE*) decrease on the training set in the consequential epochs. Although decrease of MSE means that the network is trained properly but this is not sufficient; because it might be possible that the NN begin to memorize the input patterns. As a result the generalization ability of the network is decreased, though the MSE is decreasing. Preventing this problem to occur in the training process the learning set is divided into three subsets; Training set, validation set and testing set. Sixty percent of the patterns arranged for the learning process is allocated for the training subset, twenty percent for the validation and the rest twenty percent for the testing subset. The subsets do not overlap with each others. These subsets are parts of the learning process and should not be mistaken with the testing set that is for testing the input patterns after completion of the training.

In the learning process, MSE is obtained for the three subsets at the end of each epoch. At this stage if the MSE for the training set is decreasing but is remained constant or increasing for the validation set in one or more than one epoch, then the learning of the network will come to a stop. The function of the testing set is to show the generalization ability of the network.

4.3.2   ANN Models

To investigate the mixture temperature effects on performance of ANN models and the degree of precision in prediction of slump flow, in addition to the model with 7 input neurons, a second model is developed with the same network details but 6 input neurons, in which the temperature is excluded from. Two separate ANN models are made of the same database. The results of ANN models were evaluated using the mean square error (*MSE*), mean absolute percentage error (*MAPE*) and absolute fraction of variance (R^2). The *MSE*, *MAPE* and R^2 are calculated using eqs. (1), (2) and (3).

$$\text{MSE} = \frac{1}{p} * \sum_j (t_j - O_j)^2 \ , \tag{1}$$

$$\text{MAPE} = \frac{1}{p} * \sum_j \left(\left| \frac{O_j - t_j}{O_j} \right| * 100 \right) , \tag{2}$$

$$R^2 = 1 - \left(\frac{\sum_j (t_j - O_j)^2}{\sum_j (O_j)^2} \right) , \tag{3}$$

The MSE measures the goodness-of-fit relevant to high slump values whereas R^2 is the coefficients of determination

The performance of the ANN model including mixture temperature for predicting the slump flow of testing set when the temperature is included in the inputs is illustrated in Figure 5. The results indicate that the proposed ANNs model is successful in learning the relationship between different input and the output parameters when the temperature value exists. Figure 5 illustrates that the ANNs model is capable of generalizing between inputs and the output variables with good accuracy predictions. Figure 5 illustrates the performance of the network when the temperature value is not applied to the model.

Figure 6 shows that the model is still accurate but the model generalization is lower than the first model. It can also be understood from comparing the testing sets averages between table 3 and 4. The statistical parameter for the model in which the temperature value is applied is a more precise prediction. This can be obtained when the experimental temperature value is available.

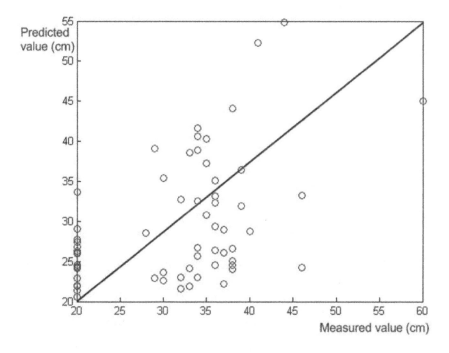

Figure. 5. Performance of training set of slump flow prediction with ANNs model when the temperature is included in the data set

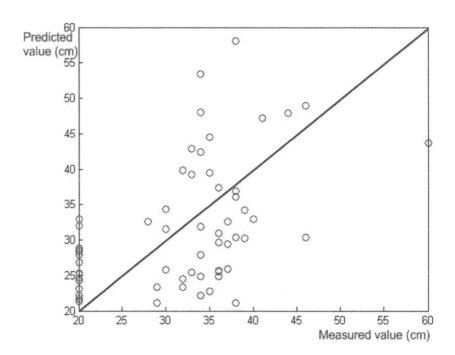

Figure. 6. Performance of testing set of slump flow prediction with ANNs model when the temperature is excluded from the data set

Table 5 and 6 show the details of the result of ANN model with and without concrete temperature value. Ten sub sets show 10 fold cross validation of the data set. Compared to table 6, table 5 presents the fact that ANN model with temperature input neurons have higher R^2 and lower MAPE and MSE.

The details of statistical parameters of training process and good training of the network are extracted at the end of the 10 learning crosses. The figures presented in appendices D.1 and D.2 indicate the state of network at the end each cross for 10 fold cross validation. Appendices D.1 shows the statistical results of the training set incorporating three subsets e.g. training, validation and testing subsets in which the temperature input value is included. Appendices D.2 shows the statistical results of the training set excluding temperature input value. Comparing absolute fraction of variance (R^2) value of training subset showed in appendices D.1 and D.2 with R^2 values presented in tables 5 and 6 illustrates a little difference. This is because the results of the NN are slightly different for different performs of the program and does not signify invalidity of the results.

Table 5 The results of slump flow model of artificial neural network including temperature value

Model	Training set			Testing set				
	Data set	Num of data	R^2	Data set	Num of data	MSE	MAPE	R^2
Model A	B,C,D, E,F,G, H,I,J	55	0.9901	A	6	47.19	19.34	0.9597
Model B	A,C,D, E,F,G, H,I,J	55	0.9968	B	6	43.31	18.63	0.9625
Model C	A,B,D, E,F,G, H,I,J	55	0.9955	C	6	87.78	26.23	0.9105
Model D	A,B,C, E,F,G, H,I,J	55	0.9930	D	6	101.01	37.98	0.8083
Model E	A,B,C, D,F,G, H,I,J	55	0.9850	E	6	54.87	20.23	0.9488

Model	Data set	Num of data	R^2	Data set	Num of data	MSE	MAPE	R^2
Model F	A,B,C, D,E,G, H,I,J	55	0.9893	F	6	19.66	10.82	0.9817
Model G	A,B,C, D,E,F, H,I,J	55	0.9993	G	6	52.04	20.06	0.9415
Model H	A,B,C, D,E,F, G,I,J	55	0.9895	H	6	40.32	19.92	0.9576
Model I	A,B,C, D,E,F, G,H,J	55	0.9558	I	6	85.12	28.40	0.9176
Model J	A,B,C, D,E,F, G,H,I	54	0.9692	J	7	83.77	21.20	0.9233
Integral testing set						**61.51**	**22.28**	**0.9311**

Table 6 The results of slump flow model of artificial neural network excluding temperature value

Model	Training set			Testing set				
	Data set	Num of data	R^2	Data set	Num of data	MSE	MAPE	R^2
Model A	B,C,D, E,F,G, H,I,J	55	0.9559	A	6	6.30	8.24	0.9939

Model B	A,C,D, E,F,G, H,I,J	55	0.9816	B	6	29.58	16.20	0.9678
Model C	A,B,D, E,F,G, H,I,J	55	0.9807	C	6	54.39	20.87	0.9440
Model D	A,B,C, E,F,G, H,I,J	55	0.9953	D	6	151.16	25.86	0.8335
Model E	A,B,C, D,F,G, H,I,J	55	0.9996	E	6	79.91	31.38	0.8829
Model F	A,B,C, D,E,G, H,I,J	55	0.9641	F	6	50.45	22.46	0.9395
Model G	A,B,C, D,E,F, H,I,J	55	0.9920	G	6	81.67	32.77	0.8878
Model H	A,B,C, D,E,F, G,I,J	55	0.9964	H	6	32.45	17.83	0.9545
Model I	A,B,C, D,E,F, G,H,J	55	0.9563	I	6	116.18	37.86	0.8369
Model J	A,B,C, D,E,F, G,H,I	54	0.9817	J	7	107.52	39.24	0.8458
Integral testing set						**70.96**	**25.27**	**0.9087**

All of the statistical values in Table 5 show that the proposed ANN model including temperature values is suitable and can be useful to predict the slump flow close to the experimental values.

4.3.3   Regression Models

To compare ANNs model results with the results of statistical techniques, all seven input parameters, are remolded with linear and nonlinear regression. To investigate the workability of fresh concrete with and without temperature effects by regression analysis, four types of regression models were applied; Enter, Forwards, Backward, And Stepwise. For each model the coefficient of determination (R^2) and mean square error (MSE) determine the precisions. In this way the best fit linear and non-linear regression models are comparatively discovered. The models correlate predicted output (slump) of the model with the target output (slump) which is obtained in the experiments. This is the same process in the NN analysis of the data. A general description of these four models is that Enter model correlates the predicted output with the measured output in a way that all the input parameters are applied for the prediction of slump in each experiment. Using this model does not omit any input value in the regression equation. Forwards model first choose the most influential input in the prediction and then add to that other important inputs. Backward first take all the inputs to have an influence on the regression equation and then eliminate inputs that are not as much effective as others. The final model used here is the Stepwise model. This model can both incorporate and eliminate inputs until it finds the influential input parameters.

 In this research SPSS version 10 was applied to determine the best fit of linear and non-linear regression.

4.3.3.1   Linear Models with temperature input value

Enter Model

In this method all the input variables are entered into the regression analysis. This method necessarily does not present the most precise results among other regression methods. Result of linear regression made for characterization of mapping among input parameters including the temperature value is given as follow:

$$Slump\ (Diameter) = 54.872 - 0.0126 * CA - 0.0268 * FA - 0.0106 * C + 0.139 * W + 0.691 * SP - 0.2 * T - 8.864 * D_{max}$$

Figure 7 presents the correlation between target and predicted slump value. The linear equation shows the relation of this correlation.

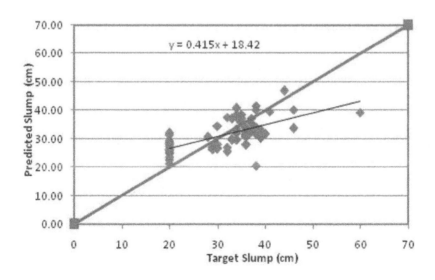

Figure. 7. Predicted slump flow values using first-order regression (Enter model) vs. measured test values (with temperature)

Regression analysis equation produced by this method present the fact that increase in coarse aggregates, cement and temperature tends to decrease the slump with different rates. On the other hand, water, superplasticizer, and the temperature coefficients are positive in above equation. Therefore increase in these inputs can increase the slump flow of fresh concrete. Most of these correlations do not vary in other methods. These corelations also could have been observed experimentally but the rate of influence of each input would be ambiguous.

The statistical parameters for this model are:

$$R^2 = 0.415$$
$$MSE = 264.4$$

The prediction precision of this method is higher than other methods of linear regressions analysis.

Forwards Model

In this method only the influential inputs are gathered in the analysis equation. Below is the equation of linear regression for Forwards model. Result of this linear regression model is presented in figure 8.

$$Slump = 13.079 + 0.105 * W + 0.693 * SP - 0.189 * T$$

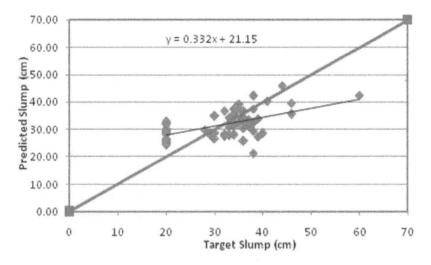

Figure. 8. Predicted slump flow values using first-order regression (Forwards model) vs. measured test values (with temperature)

Although the prediction precision of the model is not higher than Enter method, this method is useful because it only applies three input values and generates a good accuracy among linear regressions. The correlation between diameter of slumped fresh concrete and inputs chosen by this method shows that three inputs, water content, superplasticizer content and temperature dominate the workability of fresh concrete. Presence of temperature among these parameters indicates that it has an important role in determination of slump flow.

The statistical parameters for this model are:

$$R^2 = 0.332$$
$$MSE = 492.9$$

Backward Model

In this method just some of the input variables remain in the model as the most dominant inputs. Figure 9 illustrates the correlation for this method. The equation of analysis is given as follow:

$$Slump = 19.146 + 0.111 * W + 0.675 * SP - 0.194 * T - 8.839 * D_{max}$$

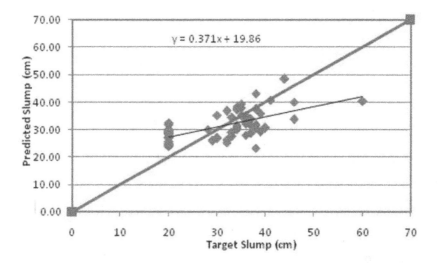

Figure. 9. Predicted slump flow values using first-order regression (Backward model) vs. measured test values (with temperature)

This method has a more precise prediction in comparison with Forwards model. Better precision of this method is related to correlating output with more input parameters e.g. maximum size of coarse aggregates (D_{max}) in comparison with Forwards model. As the Enter model verifies, Backward linear analysis shows that temperature and D_{max} of mixture are not negligible in slump prediction. Increase of both of these parameters cause the slump flow to decrease.

The statistical parameters for this model are:

$$R^2 = 0.371$$

$$MSE = 413.3$$

Stepwise Model

In this method after trying several combinations of inputs, the most influential parameters are chosen to incorporate in the analysis equation. Figure 10 shows the correlation for this method and the equation of correlation. The equation of analysis is given as follow:

$$Slump = 13.079 + 0.105 * W + 0.693 * SP - 0.189 * T$$

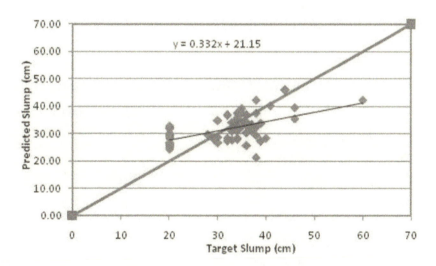

Figure. 10. Predicted slump flow values using first-order regression (Stepwise model) vs. measured test values (with temperature)

Like Forwards model, precision of this method is not higher than other two methods. This model presents that the temperature of the mixture, superplasticizer content and water content are the most effective parameters in variation of slump flow.

The statistical parameters for this model are:

$$R^2 = 0.332$$
$$MSE = 492.9$$

4.3.3.2   Linear Models without temperature input value

To investigate the influence of including temperature in the inputs of linear regression analysis, four models were developed from all experimental input parameters except the temperature value.

Enter Model

In this method all the input variables are entered into the regression analysis. This method necessarily does not present the most precise results among other regression methods. The result of linear regression for characterization of mapping among input parameters without the temperature value is given below:

$$Slump\ (Diameter) = 39.777 - 0.00752 * CA - 0.0185 * FA - 0.0266 * C + 0.164 * W + 0.709 * SP - 8.585 * D_{max}$$

Figure 11 presents the correlation between target and predicted slump value. The linear equation shows the relation of this correlation.

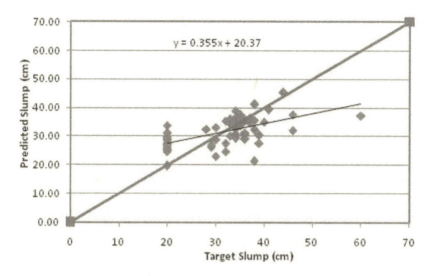

Figure. 11. Predicted slump flow values using first-order regression (Enter model) vs. measured test values (without temperature)

Regression analysis equation produced by this method present the fact that just like when the temperature is included, increase of coarse aggregates and cement has a tendency to decrease the slump with different rates, although the effect of each parameter is not as precise as linear regression including temperature input. On the other hand, coefficients of water and superplasticizer content are positive in the equation. Therefore increase in these inputs can increase the slump flow of fresh concrete. As the equation explains, the coefficients in the correlations do not strongly vary. This means that although excluding temperature inputs from the analysis reduces precision of the prediction, the correlation and the sign of coefficients remain quite fixed. These correlations also could have been observed experimentally but the rate of influence of each input would be ambiguous.

The statistical parameters for this model are:

$$R^2 = 0.355$$

$$MSE = 263.7$$

The prediction precision of this method excluding temperature inputs is also higher than other methods of linear regressions analysis.

Forwards Model

The influential inputs obtained in this model are the same as those obtained in the linear Forwards model when the temperature inputs are included. Below is the equation of linear regression for Forwards model. The result of this linear regression model is presented in figure 12.

$$Slump = 6.26 + 0.123 * W + 0.701 * SP$$

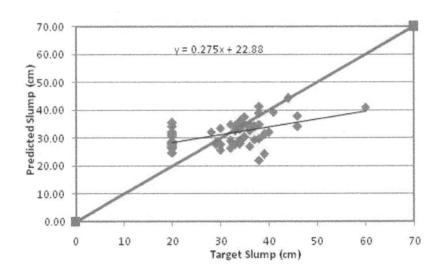

Figure. 12. Predicted slump flow values using first-order regression (Forwards model) vs. measured test values (without temperature)

The above equation indicates that when the temperature is excluded from the inputs. The key parameters of Forwards model do not vary, though in this model also the precision of the prediction shows decrease in comparison with other models.

Although the prediction precision is not higher than the previous method, this method is useful because it only applies three input values and generates a good accuracy among linear regressions. The correlation between diameter of slumped fresh concrete and inputs chosen by this method shows that three inputs, i.e., water content, superplasticizer content and temperature dominate the workability of fresh concrete. Presence of temperature among these parameters indicates that it has an important role in determination of slump flow.

The statistical parameters for this model are:

$$R^2 = 0.275$$
$$MSE = 613.5$$

Backward Model

In this method the dominant input variables are water content, superplasticizer content, and MSCA. Figure 13 illustrates the correlation for this method. The equation of analysis is given as follows:

$$Slump = 11.923 + 0.130 * W + 0.683 * SP - 8.499 * D_{max}$$

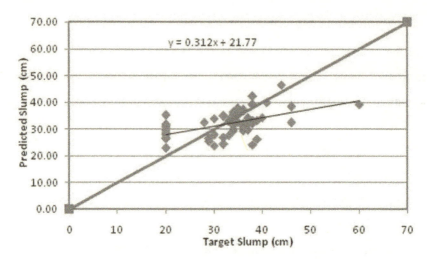

Figure. 13. Predicted slump flow values using first-order regression (Backward model) vs. measured test values (without temperature)

This method has a more precise prediction even among linear relations. Better precision of this method is related to correlating output with more input parameters, e.g. maximum size of coarse aggregates (D_{max}) in comparison with Forwards model. As the Enter model verifies, Backward linear analysis shows that temperature and D_{max} of mixture cannot be neglected in slump prediction and as the two parameters increase the slump flow decrease.

The statistical parameters for this model are:

$$R^2 = 0.312$$
$$MSE = 462.8$$

Stepwise Model

In this method the most influential parameters, that are chosen to be incorporated in the analysis equation are the same as those used in the stepwise model including temperature inputs. Figure 14 shows the correlation for this method and the equation of correlation. The equation of analysis is given as follows:

$$Slump = 6.26 + 0.123 * W + 0.701 * SP$$

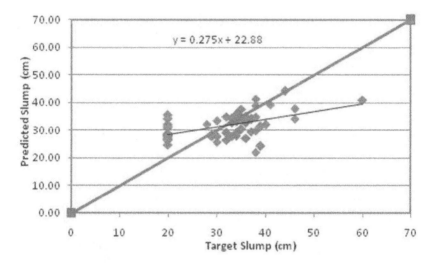

Figure. 14. Predicted slump flow values using first-order regression (Stepwise model) vs. measured test values (without temperature)

Like Forwards model, precision of this method is not higher than the other two methods. This model presents that superplasticizer content and water content are the most effective parameters in variation of slump flow even when the temperature value is removed from the inputs.

The statistical parameters for this model are:

$$R^2 = 0.275$$

$$MSE = 613.5$$

4.3.3.3 Non-Linear Models with temperature input value

In order to verify the accuracy of ANN model results, non-linear regression models were developed with and without temperature inputs. To discover the best fit non-linear regression models, four models were derived from the same data base used for linear regression and ANN models. In this way the best fit linear and non-linear regression models were comparatively discovered as well.

Enter Model

In this method all the input variables are entered into the regression analysis, either in second-order or in first-order. Among non-linear regression models of this research, this method presents the most precise results. Result of non-linear regression made for characterization of mapping among input parameters including the temperature value is given as follow:

$$Slump\,(Diameter) = 169.864 - 0.103 * CA - 0.0775 * FA - 0.566 * C + 0.128 * W + 2.345 * SP + 0.115 * T - 149.409 * D_{max} + 0.0000545 * CA^2 + 0.0775 * FA^2 - 0.566 * C^2 - 0.566 * W^2 - 0.0804 * SP^2 - 0.00759 * T^2 + 95.436\,D_{max}{}^2$$

Figure 15 presents the correlation between target and predicted slump value. The non-linear equation shows the relation of this correlation.

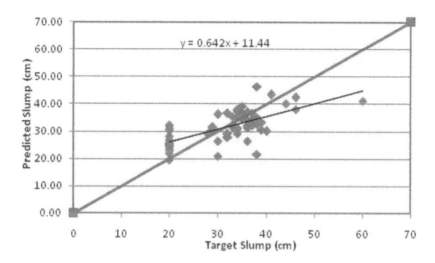

Figure. 15. Predicted slump flow values using second-order regression (Enter model) vs. measured test values (with temperature)

Regression analysis equation produced by this method can be judged directly. The equation is consisted of first order and second order inputs together. Thus the coefficients and sings of parameters do not directly indicate increasing or decreasing of slump flow, but effect of each parameter must be explained by the first and second order coefficient together. What is deduced from scrutinizing the equation is that second order coefficients regulate the first order coefficients and cannot make a big variation in the whole effect of parameters on the slump flow of fresh concrete. In this way they call for greater precision to be produced for the non-linear regressions rather than linear regression of the same models. Finally the equation generated by this model presents the fact that increase of coarse aggregates, cement and temperature has a tendency to decrease the slump with different rates. On the other hand, water, superplasticizer, and the temperature coefficients show that increase in these inputs can increase the slump flow of fresh concrete.

The statistical parameters for this model are:

$$R^2 = 0.642$$

$$MSE = 204.3$$

The prediction precision of this method is higher than other methods of non-linear regressions analysis.

Forwards Model

In this method the influential inputs gathered in the analysis equation are superplasticizer content, water content and the temperature. Below is the equation of non-linear regression for Forwards model. Result of this non-linear regression model is presented in figure 16.

$Slump\ (Diameter) = 12.759 + 2.673 * SP - 0.566 * W^2 - 0.101 * SP^2 - 0.00432 * T^2$

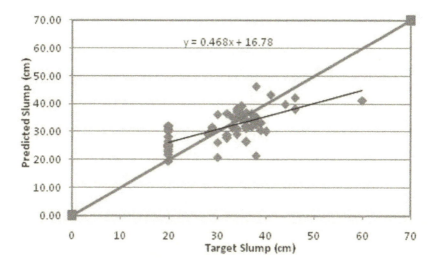

Figure. 16. Predicted slump flow values using second-order regression (Forwards model) vs. measured test values (with temperature)

In this method also the coefficients of parameters do not directly indicate the correlation between input and output values. Although the prediction precision of the model is not higher than Enter method, this method is useful because it only applies three input values and generates a good accuracy among non-linear regressions. The correlation between diameter of slumped fresh concrete and inputs chosen by this method shows that three inputs, water content, superplasticizer content and temperature dominate the workability of fresh concrete in this method too. Presence of temperature among these parameters indicates it has an important role in determination of slump flow.

The statistical parameters for this model are:

$$R^2 = 0.468$$

$$MSE = 521.7$$

Backward Model

In this method most of the input variables correlate with the output with different orders. Figure 17 illustrates the correlation for this method. The equation of analysis is given as follow:

$$Slump\ (Diameter) = 139.207 - 0.49 * C + 0.126 * W + 2.495 * SP - 142.02 * D_{max} + 0.0007 * C^2 - 0.0892 * SP^2 - 0.00516 * T^2 + 90.462\ D_{max}{}^2$$

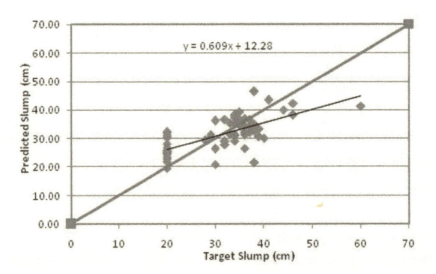

Figure. 17. Predicted slump flow values using second-order regression (Backward model) vs. measured test values (with temperature)

Non-linear regression for Backward model has a more precise prediction in comparison with Forwards model. Higher precision of this method is related to correlating output with more input parameters e.g. maximum size of coarse aggregates (D_{max}) and cement content in comparison with Forwards model. As the Enter model verifies, Backward linear analysis shows that temperature and D_{max} of mixture are not negligible in slump prediction and as these parameters increase the slump flow decrease.

The statistical parameters for this model are:

$$R^2 = 0.609$$
$$MSE = 339.5$$

Stepwise Model

In this method after trying several combinations of inputs, the most influential parameters are chosen to incorporate in the analysis equation. Figure 18 shows the correlation for this method and the equation of correlation. The equation of analysis is given as follow:

$Slump\ (Diameter) = 12.759 + 2.673 * SP + 0.00033 * W^2 - 0.101 * SP^2 - 0.00432 * T^2$

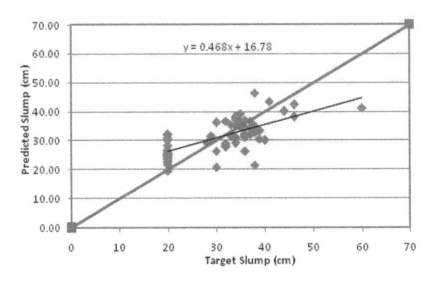

Figure. 18. Predicted slump flow values using second-order regression (Stepwise model) vs. measured test values (with temperature)

Like Forwards model, precision of this method is not higher than other two methods. This model shows that the temperature of the mixture, superplasticizer content and water content are the most effective parameters in variation of slump flow.

The statistical parameters for this model are:

$$R^2 = 0.468$$
$$MSE = 521.7$$

4.3.3.4   Non-Linear Models without temperature input value

To investigate the non-linear analysis with and without experimental temperature value in the inputs, four models are developed and the results are presented as follow.

Enter Model

Result of linear regression made for characterization of mapping among input parameters without the temperature value is given as follow:

$$Slump\ (Diameter) = 161.463 - 0.121 * CA + 0.0974 * FA - 0.637 * C + 0.13 * W + 2.32 *\ SP - 122.679\ * D_{max} + 0.0000691 * CA^2 - 0.0000544 * FA^2 + 0.000902 * C^2 + 0.0000837 * W^2 - 0.0795 * SP^2 + 78.234\ D_{max}{}^2$$

Figure 19 presents the correlation between target and predicted slump value. The non-linear equation shows the relation of this correlation.

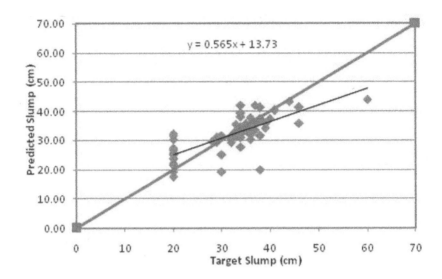

Figure. 19. Predicted slump flow values using second-order regression (Enter model) vs. measured test values (without temperature)

Regression analysis equation produced by this method present the fact that just like when temperature is included, increase of coarse aggregates and cement has a tendency to decrease the slump with different rates, although this method does not measure the effect of each parameter as precisely as non-linear regression.

The statistical parameters for this model are:

$$R^2 = 0.565$$
$$MSE = 209.8$$

The prediction precision of this method excluding temperature inputs is higher than other methods of linear regressions analysis.

Forwards Model

The influential inputs obtained in this model are the same as those obtained in the non-linear Forwards model when the temperature inputs are included. Below is the equation of linear regression for Forwards model. Result of this linear regression model is presented in figure 20.

$$Slump\ (Diameter) = 9.351 + 2.641 * SP + 0.000377 * W^2 - 0.0997 * SP^2$$

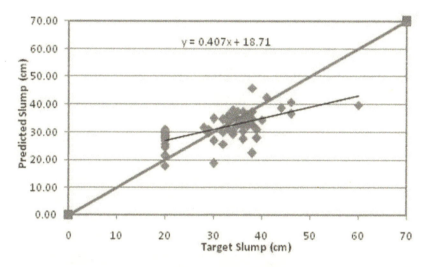

Figure. 20. Predicted slump flow values using second-order regression (Forwards model) vs. measured test values (without temperature)

The above equation indicates that when temperature is removed from the inputs of non-linear regression, among the key parameters of Forwards model, only the coefficient of water content changes to positive. This shows the fact that in different temperatures the water content of the mixture does not have a specific effect on the workability of fresh concrete.

The statistical parameters for this model are:

$$R^2 = 0.407$$
$$MSE = 604.5$$

Backward Model

In this method most of the input variables correlate with the output with different orders. Figure 21 illustrates the correlation for this method. The equation of analysis is given as follow:

$$Slump\ (Diameter) = 178.847 - 0.086 * CA - 0.586 * C + 0.151 * W + 2.274 * SP - 120.524 * D_{max} + 0.0000498 * CA^2 + 0.000826 * C^2 - 0.0773 * SP^2 + 76.654\ D_{max}{}^2$$

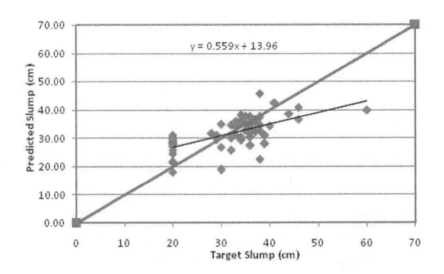

Figure. 21. Predicted slump flow values using second-order regression (Backward model) vs. measured test values (without temperature)

Non-linear regression excluding temperature value for Backward model has a more precise prediction in comparison with Forwards model. Higher precision of this method is related to correlating output with more input parameters e.g. maximum size of coarse aggregates (D_{max}) and cement content in comparison with Forwards model.

The statistical parameters for this model are:

$$R^2 = 0.559$$
$$MSE = 276.6$$

Stepwise Model

In this method after trying several combinations of inputs, the most influential parameters are chosen to be incorporated in the analysis equation. Figure 22 shows the correlation for this method and the equation of correlation. The equation of analysis is given as follow:

$Slump\ (Diameter) = 9.351 + 2.641 * SP + 0.000377 * W^2 - 0.0997 * SP^2$

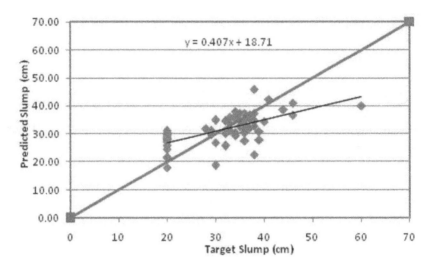

Figure. 22. Predicted slump flow values using second-order regression (Stepwise model) vs. measured test values (without temperature)

Like Forwards model, precision of this method is not higher than other two methods. This model presents that, superplasticizer content and water content are the most effective parameters in variation of slump flow.

The statistical parameters for this model are:

$$R^2 = 0.407$$
$$MSE = 604.5$$

4.3.4   Comparison between Regression Models

To observe the effect of temperature input neurons on slump flow prediction of fresh concrete, linear and non-linear regression models are compared. Figure 23 illustrates the linear prediction of slump value with and without temperature. Figure 24 respectively shows the non-linear regression prediction of slump value with and without temperature. Figure 23 and Figure 24 show that when temperature is applied, the accuracy of the prediction increases. The statistical parameters and rate of increments for these models are presented in table 7. The result of non-linear regressions verifies the accuracy of linear models. From all the regression models it is concluded that temperature has a significant influence on slump flow and its variation can enormously change the workability of fresh concrete.

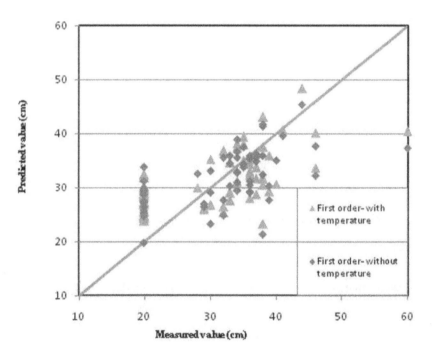

Figure. 23. Predicted slump flow values using first-order regression vs. measured test values with and without including temperature

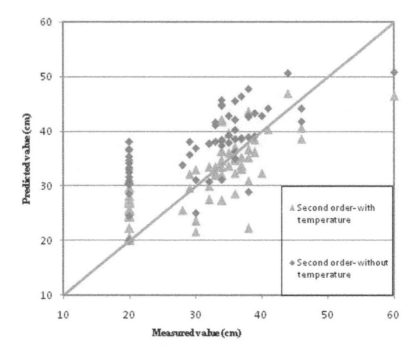

Figure. 24. Predicted slump flow values using second-order regression vs. measured test values with and without including temperature

4.3.5   Comparison between ANN Model and Regression Models

The comparison between ANN and regression models shown in table 7 demonstrates that the neural network models (R^2=0.931) are supported better by the experimental data than the non-linear (R^2=0.642) and linear (R^2=0.415) regression models. It is also shown that although recording a wide range of temperatures for each experiment is not as simple as other variables, it can provide higher degree of accuracy for the model. Table 7 also shows that in all the methods the prediction precision increases with including temperature as an inputs.

Table 7 Statistical parameters of neural networks and regression models

Statistical parameters (With Temperature)	ANNs model	Linear regression model (Enter)	Non-linear regression model (Enter)
	Integral Testing set	Integral Testing set	Integral Testing set
MSE	61.51	264.40	204.3
R^2	0.9311	0.415	0.642
Statistical parameters (Without Temperature)	ANNs model	Linear regression model (Enter)	Non-linear regression model (Enter)
	Integral Testing set	Integral Testing set	Integral Testing set
MSE	70.96	263.77	209.89
R^2	0.9087	0.355	0.565

CHAPTER FIVE

ANALYSIS AND DISCUSSION

5.1   Outline

This section presents the discussion about the results from the experimental research plan presented in Chapter 4. The effect of input parameters especially temperature on the slump flow of fresh concrete will be discussed. To assess the relationships, statistical analyses were performed between the NN predicted slump values and experimental input values.

• Effect of temperature of fresh concrete on the slump flow
• Effect of each input parameter on the slump flow
• Interaction of input parameters with each other and their total effect on slump flow
It must be mentioned that the correlations may change regarding to materials, types of experiments and mixtures.

5.2   Effect of Temperature on Slump Flow

The main scope of this study was to focus on effect of 'As Placed' fresh concrete temperature on the slump flow. Figure 25 shows the correlation between fresh concrete temperature and slump flow. This is a NN prediction of slump flow correlated with temperature. The prediction precision is discussed in chapter 4. As shown in figure 25 the slump flow shows constant decrease when temperature increases. The rapid rate of evaporation of moisture and increased rate of slump loss at elevated temperatures are the basic reasons for the slump flow decrease. This means that at higher temperatures initial shear stress which is measured qualitatively with the slump flow test increases and produces harsher concretes with lower slumps.

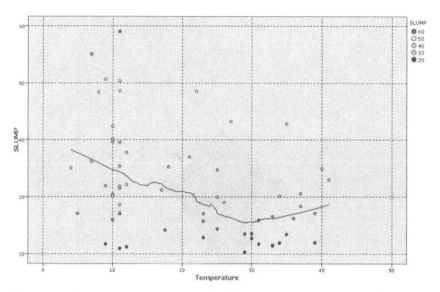

Figure. 25. The correlation between fresh concrete temperature and predicted slump flow

5.3   Correlation of Temperature with Other Parameters

The correlation of concrete temperature with slump flow is not constantly the way presented in figure 25. Other parameters influence this correlation. The most variant parameter is the water content. Figure 26 shows the correlation of slump flow, temperature and water content together. The distribution of the surface shows that increase with water in one hand and decrease with temperature on the other cause slump flow increase. In addition, it shows that at higher water contents the rate of slump loss is higher than that of lower water contents. Moreover, the water increment at low temperatures can more severely increase slump rather than high temperatures. This represents the fact that when comparing these three parameters regardless of others, the most dominant factor defect of slump loss is the lack of water.

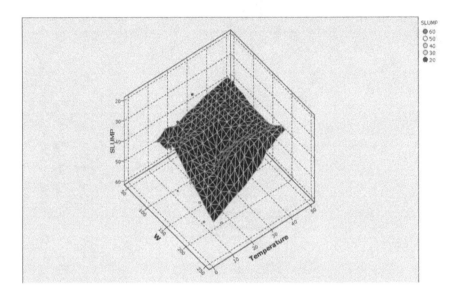

Figure. 26. The correlation between fresh concrete temperature, water and predicted slump flow

Another parameter is the maximum size of coarse aggregates (MSCA). Figure 27 shows the correlation of slump flow, temperature and MSCA. The figure illustrates that decrease of maximum size of coarse aggregates significantly increase the slump. It is also shown that when finer maximum size of coarse aggregates is chosen, more fluid concrete is achieved. This allows temperature to show its effect more clearly. Intensified effect of temperature on slump at low MSCA in comparison with high MSCA may be related to the experiments' arrangements or it may be related to the fact that at low slumps, heat gained from internal friction during mixing of concrete will be greater and helps increasing temperature.

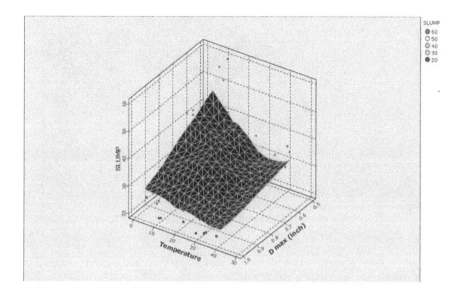

Figure. 27. The correlation between fresh concrete temperature, MSCA and predicted slump flow

Like water content, superplasticizer content can affect the correlation of temperature and slump flow. Figure 28 shows this correlation. Temperature effect on slump can be seen with the same general surface shape. According to figure 28 the superplasticizer used in the mixture can replace a slump of 25 cm (diameter) to over 40 cm at temperatures between 0 to 10 degree of centigrade and a slump of 20 cm to over 35 cm at temperatures between 30 to 40 degree of centigrade. The increase of slump with superplasticizer does not differ very much for both hot and cold temperatures.

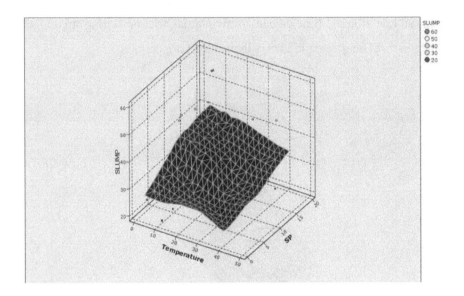

Figure. 28. The correlation between fresh concrete temperature, superplasticizer and predicted slump flow

Effect of fine aggregates on the correlation of temperature with slump is different from liquid-like ingredients. As noted in the literature about the effect of fine aggregates on slump flow, figure 29 also shows that increase of finer aggregates will increase the slump flow and does not produce a distinct effect on the correlation of the temperature with slump in this study.

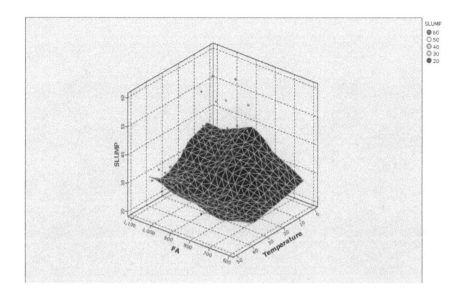

Figure. 29. The correlation between fresh concrete temperature, fine aggregates and predicted slump flow

The other parameter relating to the temperature that cause varied rate of heat-of-hydration is the cement content. Figure 30 shows the correlation of slump flow, temperature and cement content. Figure 30 shows that the cement content does not cause variations on the slump flow of fresh concrete directly but at high cement rates the increased heat-of-hydration will increase the temperature and the rate of slump loss. Therefore as shown in figure 30 the slump flow at high cement contents is lower than slump flow at low cement contents at a constant temperature. This is also related to production of harsher and slower moving mixtures at high cement content concretes.

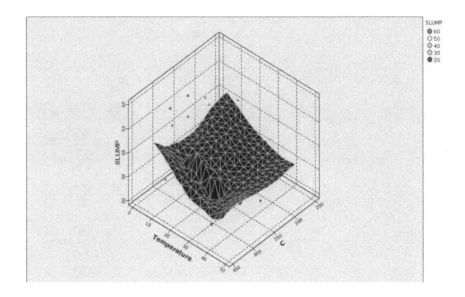

Figure. 30. The correlation between fresh concrete temperatures, cement content and predicted slump flow

5.4 Effects of Other Parameters

An overview of the effect of some ingredients was given in section 5.2. In this section the detailed independent effect of the most important parameters and their correlation with the slump flow is discussed. Figure 31 shows the correlation between water content and predicted slump flow. Figure 32 shows the correlation between superplasticizer content and predicted slump flow. Figure 31 shows increased slump flow with increasing water content. Two zero slope shown in figure 31 is probably pertaining to the types of experiments and mixtures. The general interpretation of figure 31 is that increase in water content will greatly increase the slump flow. This increase is much more rapid at water content over 200 kg/m^3; Because of at water contents over 200 kg/m^3, the W/C ratio of the mixtures has mostly been over 0.45. Figure 32 shows the correlation between the superplasticizer content and predicted slump flow. It shows that the use of superplasticizer in low amount has not increased the slump flow. Experiments showed that it is related to the type of superplasticizer and its power to produce a fluid mixture. According to figure 32 at higher superplasticizer content a sharp and then a slight increase of slump flow with superplasticizer content occur. This demonstrates that the optimum use of this superplasticizer content presented in appendix B is in the range of 5 to 10 kg/m^3. Comparing figures 31 and 32 reveals that, for this study increase of water content is much more effective in increasing slump rather than increase of superplasticizer content.

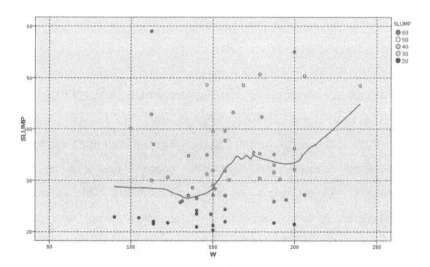

Figure. 31. The correlation between fresh water content and predicted slump flow

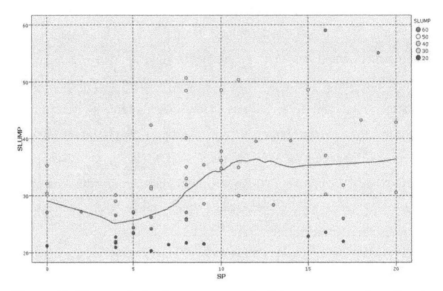

Figure. 32. The correlation between superplasticizer content and predicted slump flow

To discuss the effect of water and superplasticizer content together on the slump flow figure 33 was extracted from the results. Figure 33 demonstrates that superplasticizer can transform a slump of 35 centimeters to around 60 centimeters at high water contents and a slump of 20 centimeter to 35 centimeter at low water contents. This means that superplasticizer is soluble at water so it increases slump better at high water content ranges. Besides, the optimum dosage of superplasticizer content is shown in figure 32.

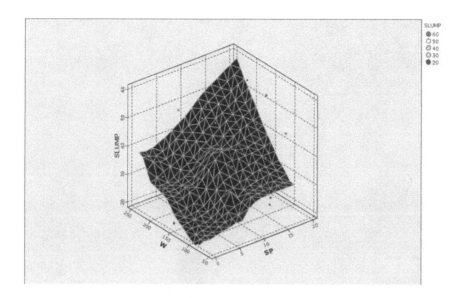

Figure. 33. The correlation between water content, superplasticizer content and predicted slump flow

Increase of coarse aggregates generally cause decrease of slump flow, although the slump depends on other factors such as aggregates water absorption, angularity, etc. Mild slope shown in figure 34 presents the expected slump decrease corresponding to coarse aggregate increase.

CA-Slump

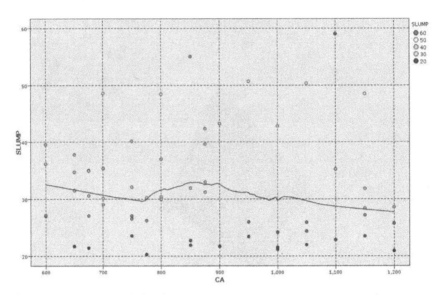

Figure. 34. The correlation between coarse aggregates content and
predicted slump flow

To investigate the effect of coarse aggregates and superplasticizer on
the slump flow, figure 34 was derived from the results. The figure shows
that coarse aggregates do not affect the correlation between
superplasticizer and slump flow. Besides, it validates the effect of
superplasticizer content on the slump flow which was discussed before.

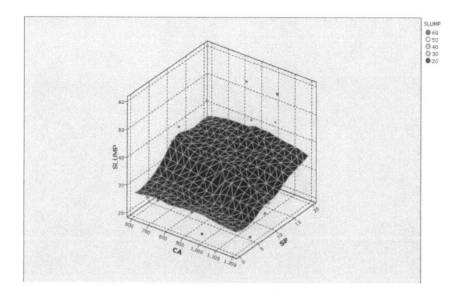

Figure. 35. The correlation between coarse aggregates, superplasticizer content and predicted slump flow

The maximum size of coarse aggregates effect on the correlation between superplasticizer and slump flow is also investigated. In this study, the other efficient parameter in this correlation is MSCA. Figure 36 demonstrates that use of coarser aggregates especially at high superplasticizer contents cause decrease in the slump flow of fresh concrete.

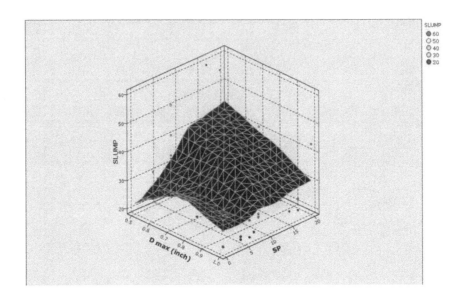

Figure. 36. The correlation between MSCA, superplasticizer content and predicted slump flow

A similar investigation of the effect of MSCA is generated in figure 36. This figure shows the correlation between MSCA, superplasticizer and predicted slump flow. The sharp gradient of slump decrease at high water contents presented in figure 37 shows the same effect the coarser aggregates have on the correlation of superplasticizer and slump.

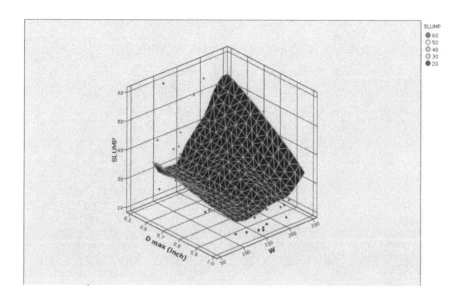

Figure. 37. The correlation between MSCA, Water content and predicted slump flow

As mentioned in the literature, increase of fine aggregates generally cause increase of slump flow, although the slump depends on other factors such as fineness. Mild slope shown in figure 34 presents the expected slump increase corresponding to fine aggregate increase.

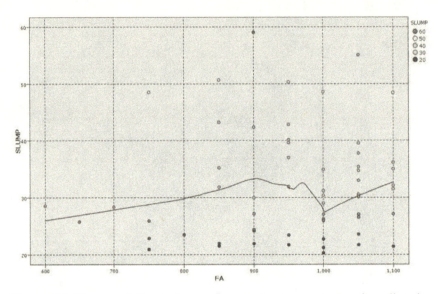

Figure. 38. The correlation between fine aggregates content and predicted slump flow

Water content and water absorption are greatly dependent on fine aggregates content. Thus the total effect of these two parameters on the slump flow is of great importance.

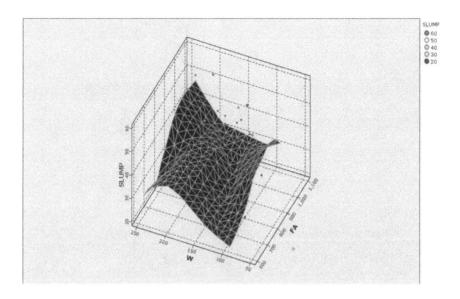

Figure. 39. The correlation between fine aggregates, water content and predicted slump flow

CHAPTER SIX

COCLUSION AND
RECOMMENDATIONS

6.1   Summary

The Primary objective of this study was to evaluate the influence of inclusion of concrete temperature in ANN models for predicting the workability determined by the standard slump cone test ASTM C 143. The secondary objective of this study was to investigate the interaction between temperature and other factors in order to determine a specific slump, and the correlation between ingredients and slump when temperature is applied in the network. This study investigated the slump flow property of conventional and high strength concretes. Measured workability was obtained in the laboratory.

The laboratory phase of this study investigated slump flow property of sixty one mixtures represented in appendix C.1. The mixture designs are calculated to make scattered data.

Neural network analyses were performed to assess the significance of the following relationships: Fresh concrete temperature input value and neural network prediction precision; Neural network analyses and statistical analyses; Fresh concrete temperature and slump flow amount (ASTM C 143); Concrete ingredients and slump flow amount.

6.2   Conclusion

Because of the broad variation in compositions and characteristics of concrete materials, it is usually difficult to provide an accurate model that is suitable to predict the workability of most concretes, especially for models which include temperature effects. The proposed methodology provides a guideline to model complex material behaviors using only a limited amount of experimental data. This study led to the following conclusions:

1. The concrete temperature cannot be ignored when predicting concrete workability using ANN. It has a factual impact on accuracy of the network prediction. When the temperature is included (0.9311), the coefficient of determination (R^2) of slump flow model of ANN is greater than the time temperature is not included (0.9087). Therefore, the slump flow model which includes parameter of temperature is more accurate than the model which does not include temperature (Figures 5, 6).

2. The best coefficient of determination (R^2) of slump flow second-order regression model when the temperature is included (0.642) is greater than the time temperature is not included (0.565). Therefore, the slump flow model including parameter of temperature is more accurate than the model which does not include temperature (Figures 15, 19).

3. The coefficient of determination R^2 of slump flow model of ANN is greater than that of second-order regression for all regression models (enter, forward, backward, and stepwise) in both models (with and without temperature). Therefore, the slump flow model based on ANN is much more accurate than the model based on the regression analysis. The second order models also verify the accuracy of the ANN model results (Table 7).

4. The coefficient of determination R^2 of slump flow second-order regression model is much greater than that of first-order regression for all regression models (enter, forward, backward, stepwise) in both models (with and without temperature). Therefore, the slump flow model based on

ANN is much more accurate than the model based on the regression analysis (Table 7).

5. By using response trace plots, it is found that the slump flow decreases sharply as the concrete temperature content increases; It decreases more sharply at high superplasticizer contents rather than low contents (Figures 25 and 28).

6. By using response trace plots, it is also found that the slump flow decreases more sharply at bigger MSCA (Figure 27).

7. The effects of two parameters, e.i. the water content and the MSCA can cause significant variation in slump-temperature interaction (Figures 26 and 27).

8. The equations derived from regression models, presented in section 4.3.3, and the figures shown in chapter five can be of use in concrete projects. by using these equations and figures, quality control sections of concrete projects that deal with concrete placing with different as-placed temperatures and look for the desirable workability can obtain an initial approximate (because of the difference in property of materials) estimation of the slump flow of concrete. Also, the required temperature of fresh concrete, when the slump flow is significant and the other parameters of the equations are constant is predictable as well.

6.3   Recommendations for Future Researches

The inability of laboratory-based models to predict the full range of field results may be related to the type of aggregates, cement, additives and admixtures used in the field mixtures. The laboratory mixtures have a lower w/c ratio compared to the field mixtures, which may also play a role in the predictive capabilities. The following are suggestions for future researches:

A. Investigation on the effect of fresh concrete temperature on the rheological parameters, e.g. initial shear stress and viscosity of SCC

B. Investigation on the effect of fresh concrete temperature on the durability and strength of HSC

C. Laboratory investigations vs. field investigation; the effect of fresh concrete rheological parameters on the mechanical properties of hardened concrete

D. Investigation on the effect of 28 days curing temperature on the strength development of concrete

References

1. J. Ortiz, A. Aguado, L. Agullo, T. Garcı́a, "Influence of environmental temperatures on the concrete compressive strength: Simulation of hot and cold weather conditions". Cement and Concrete Research 35 (2005) 1970 – 1979

2. A.M. Neville, Properties of Concrete, fourth edition, Pearson Education Limited, England, 1999.

3. M. Mouret, A. Bascoul, G. Escadeillas, Cement and Concrete Research 27 (3) (1997) 345– 357.

4. K.A. Soudki, E.F. El-Salakawy, N.B. Elkum, Full factorial optimization of concrete mix design for hot climates, Journal of Materials in Civil Engineering 13 (6) (2001) 427–433.

5. Hampton, J. S., 1981, "Extended Workability of Concrete Containing High-Range Water-Reducing Admixtures in Hot Weather," *Developments in the Use of Superplasticizers*, SP-68, V. M. Malhotra, ed., American Concrete Institute, Farmington Hills, Mich., pp. 409-422.

6. Mittelacher, M., 1985, "Effect of Hot Weather Conditions on the Strength Performance of Set-Retarded Field Concrete," *Temperature Effects on Concrete,* STP 858, ASTM, Philadelphia, pp. 88-106.

7. Robert W. Previte, "Concrete Slump Loss," *ACI Journal,* ACI, August 1977, pp. 361-367.

8. Seung-Chang Lee, Prediction of concrete strength using artificial neural networks. Eng Structure 2003; 25: 849–57

9. Yeh IC. Modeling concrete strength with augment-neuron networks. J Mater Civil Eng 1998;10(4):263–8.

10. Brown DA, Murthy PLN, Berke L. Computational simulation of composite ply micromechanics using artificial neural networks. Microcomput Civil Eng 1991;6:87–97.

11. Bishop CM. Neural networks for pattern recognition. Oxford: Clarendon Press; 1995.

12. Bai J, Wild S, Ware JA, Sabir BB. Using neural networks to predict workability of concrete incorporating metakaolin and fly ash. Adv Eng Software 2003;34(11–12):663–9.

13. G. H. Tattersall, The workability of concrete, A viewpoint Publication, PCA 1976.

14. Glossary of Standard Rheological Terms, British Standard Institution BS 5168:1975.

15. S. H. Kosmatka and W. C. Panarese, Design and Control of Concrete Mixtures, PCA 1994.

16. US Department of Transportation, Federal Highway Administration, (APRIL 2001). Portland-Cement Concrete Rheology and Workability: Final Report, Publication No. Fhwa-Rd-00-025.

17. Scanlon, J. M. (1994). "Factors influencing concrete workability." P. Klieger, and J. F. Lamond, ed., *Significance of tests properties of concrete and concrete-making*. American Society for Testing and Materials, STP 169C, Philadelphia, PA.

18. Bartos, P. (1993a). "Workability of special concrete mixes," *Materials and Structures* 26(155), 50.

19. Bartos, P. (1993b). "Special concretes: Workability and mixing." *Proc. Intl RILEM Workshop*. TA416R5P, No. 24, E&FN Spon, London.

20. Glanville, W. R., Collins, A. R., and Mathews, D. D. (1947). "The grading of aggregate and workability of concrete," *Road Research Tech. Paper* (5), H. M. Stationary Office, London.

21. Malek, R. I. A. and Roy, D. M. (1992). "Effects of superplasticizen on the workability of concrete as evident from apparent viscosity, yield stress and zeta-potential." *Materials Research SocietySymposium, Proceeding on Flow and Microstructure of Dense Suspension*, published by Materials Research Society, Pittsburgh, PA.

22. Cordon, W. A. (1955). "Entrained air — a factor in the design of concrete mixes," *Journal of ACI, Proceedings* 51, 881, May.

23. Punkki, J., Golaszewski, J., and Gprv, O. (1996). "Workability loss of high-strength concrete," *ACI Materials Journal* 93(5), 427.

24. Kucharska, L., and Moczko, M. (1994). "Influence of silica fume on the rheological properties of the matrices of high-performance," *Adv Cem Res* 6, 139-145.

25. Male, P. L. (1993). "Workability and mixing of high performance microsilica concrete, Special Concretes: Workability Mixing." *Proceedings, International RILEM Workshop.* P. J. M. Bartos, ed., paper 19, Paisley, Scotland, 177-180.

26. I-Cheng Yeh, Modeling slump flow of concrete using second-order regressions and artificial neural networks. Cement & Con Composites 2007; 29: 474–80.

27. Kasperkiewicz J, Racz J, Dubrawski A. HPC strength prediction using artificial neural network. J Comput Civil Eng 1995;9(4):279–84.

28. Nehdi M, El Chabib H, El Naggar MH. Predicting performance of self-compacting concrete mixtures using artificial neural networks. ACI Mater J 2001;98(5):394–401.

29. Yeh IC. Exploring concrete slump model using artificial neural networks. J Comput Civil Eng 2006;20(3):217–21.

30. Ferraris, C. F. (1999). "Measurement of rheological properties of high performance concrete: State of the art report," Journal of Research of the National Institute of Standards and Technology, J. Res. Natl. Inst. Stand. Technol. 104, 461.

VITA

Mohamadreza Moini was born in Qom, Iran in 1986. He attended high school in mathematics and physics. In 2005 Moini enrolled in Civil Engineering graduate program at University of Qom. In spring of 2009 he received his Bachelor Degree in Civil Engineering. Afterwards, beside his researches, he worked as supervisor or manager of construction projects. His research interests are basically in the fields of: Concrete Rheology, Concrete additives, polymer concrete and materials, Seismic Design of Structures, Structure and Architecture. He has published many papers on the above-cited issues and has participated in many scientific and industrial committees. He has participated in compiling national standards on safety and pre-cast concrete manholes for the Institute of Standards and Industrial Research of Iran. He pursues his researches in materials and concrete technology, cooperating with research institutes.

Contact: Moini.Mohamadreza@gmail.com

Appendices

Appendix A. The fine and coarse aggregates properties for the laboratory mixtures.

Superplasticizer	ASTM C 494	$\rho(\frac{gr}{cm^3})$	PH
Properties	Type F	1.2	6 - 8.5

Grading requirements for coarse aggregates according to ASTM C-33: The CA used in this study is almost between standard limitations.

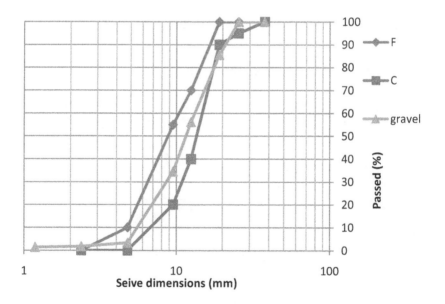

Grading requirements for fine aggregates according to ASTM C-33: The FA used in this study is almost between standard limitations.

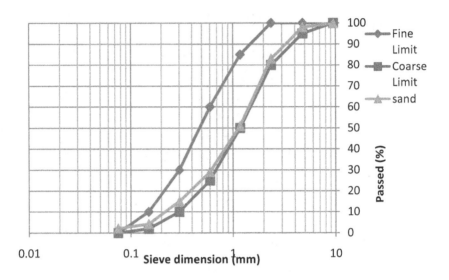

Appendix B. Properties of the chemical admixture

Superplasticizer	ASTM C 494	$\rho(\dfrac{gr}{cm^3})$	PH
Properties	Type F	1.2	6 - 8.5

Appendix C.1. Sixty-one mixture proportions for the laboratory tests according to their numbers

Ingredient	1	2	3	4	5	6
CA	600	600	600	600	650	650
FA	1100	1100	1050	1050	1100	1050
C	400	375	350	300	375	375
W	200	150	157.5	150	187.5	187.5
SP	10	5	5	12	6	4
T of mix	7	39	11	11	11	31
D max (inch)	3/4	1/2	3/4	1/2	3/4	3/4

Ingredient	7	8	9	10	11	12
CA	650	650	675	675	675	675
FA	1050	1050	1100	1100	1050	1050
C	350	300	400	375	350	350
W	157.5	135	200	187.5	157.5	122.5
SP	10	10	7	8	0	20
T of mix	12	25	33	4	5	37
D max (inch)	1	1/2	1	3/4	1	3/4

Ingredient	13	14	15	16	17	18
CA	675	700	700	700	700	750
FA	1000	1050	1050	1000	1000	1100
C	325	350	400	300	325	400
W	146.25	175	160	150	146.25	200
SP	11	9	4	4	15	0
T of mix	40	11	34	26	11	12
D max (inch)	1/2	1	1/2	3/4	3/4	1

ngredient	19	20	21	22	23	24
CA	750	750	750	750	775	775
FA	1050	1050	1000	950	1050	1000
C	350	350	300	250	425	325
W	140	140	135	100	191.25	195
SP	16	4	8	8	16	6
T of mix	30	33	23	10	10	36
D max (inch)	1/2	3/4	1	1/2	1	1

Ingredient	25	26	27	28	29	30
CA	775	800	800	800	800	850
FA	1000	1100	1000	950	900	1050
C	300	400	325	325	250	400
W	150	240	178.75	113.75	112.5	200
SP	6	8	0	16	11	19
T of mix	29	8	10	21	25	7
D max (inch)	1	3/4	1	3/4	3/4	1/2

Ingredient	31	32	33	34	35	36
CA	850	850	850	875	875	875
FA	1000	950	900	1050	1000	950
C	350	300	350	375	325	350
W	105	150	157.5	187.5	146.25	157.5
SP	4	8	4	8	6	14
T of mix	30	9	34	41	17	10
D max (inch)	1/2	1/2	1	1/2	1	3/4

Ingredient	37	38	39	40	41	42
CA	875	900	900	950	950	950
FA	900	950	850	1000	950	850
C	300	350	325	375	425	325
W	180	122.5	162.5	131.25	148.75	178.75
SP	6	8	18	17	5	8
T of mix	10	9	27	10	35	9
D max (inch)	3/4	3/4	3/4	1	3/4	1/2

Ingredient	43	44	45	46	47	48
CA	1000	1000	1000	1000	1050	1050
FA	1000	950	900	850	950	900
C	375	375	350	325	375	350
W	150	112.5	140	113.75	206.25	157.5
SP	0	20	6	9	11	5
T of mix	12	35	17.5	33	11	25
D max (inch)	1	1	3/4	3/4	3/4	1/2

Ingredient	49	50	51	52	53	54
CA	1050	1050	1100	1100	1100	1150
FA	850	750	900	850	750	900
C	325	375	375	325	300	375
W	113.75	187.5	112.5	178.75	90	206.25
SP	17	8	16	0	15	2
T of mix	39	31	11	18	23	11
D max (inch)	1	1	1/2	3/4	1	1

Ingredient	55	56	57	58	59	60	61
CA	1150	1150	1150	1150	1200	1200	1200
FA	850	800	750	700	750	650	600
C	350	350	375	275	350	325	275
W	157.5	140	168.75	151.25	140	130	137.5
SP	17	5	10	13	4	8	9
T of mix	11	29	22	37	11	23	11
D max (inch)	3/4	1/2	3/4	1	1	1/2	3/4

Appendix C.2. Slump flow results for laboratory concrete mixtures

Test number	1	2	3	4	5	6
Slump flow Diameter (cm)	46	20	34	34	30	28

Test number	7	8	9	10	11	12
Slump flow Diameter (cm)	36	38	34	35	30	20

Test number	13	14	15	16	17	18
Slump flow Diameter (cm)	40	46	20	30	32	20

Test number	19	20	21	22	23	24
Slump flow Diameter (cm)	36	20	29	39	60	33

Test number	25	26	27	28	29	30
Slump flow Diameter (cm)	32	38	20	20	33	44

Test number	31	32	33	34	35	36
Slump flow Diameter (cm)	38	35	20	36	20	34

Appendix C.2. Slump flow results for laboratory concrete mixtures

Test number	37	38	39	40	41	42
Slump flow Diameter (cm)	33	20	38	36	20	34

Test number	43	44	45	46	47	48
Slump flow Diameter (cm)	20	20	20	20	41	34

Test number	49	50	51	52	53	54
Slump flow Diameter (cm)	20	38	39	20	29	36

Test number	55	56	57	58	59	60	61
Slump flow Diameter (cm)	35	36	37	37	32	34	37

Appendix D.1. Detailed statistical results of the 10 fold training set incorporating training subset, validation subset and testing subset including temperature input.

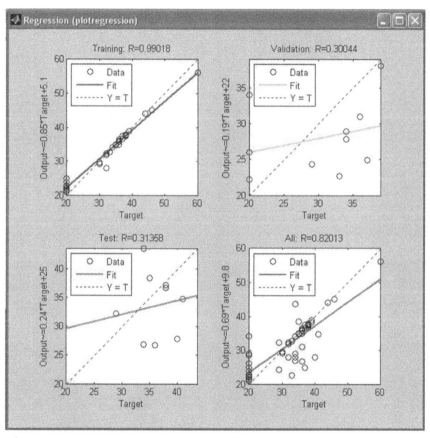

1st training set

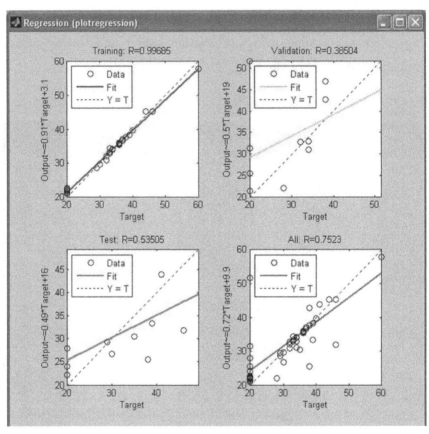

2nd training set

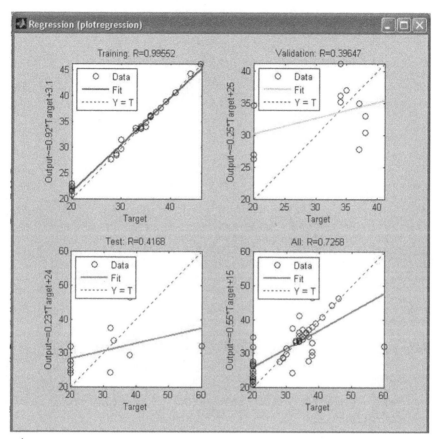

3rd training set

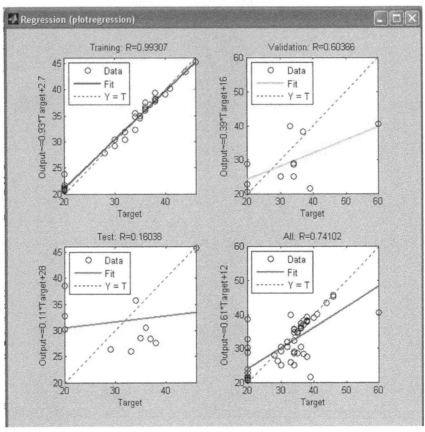

4th training set

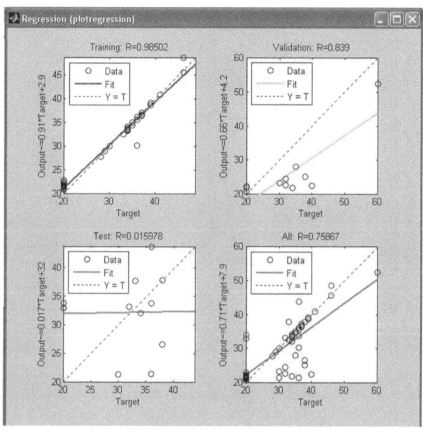

5th training set

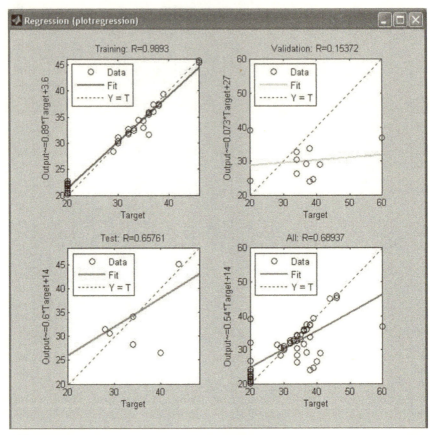

6th training set

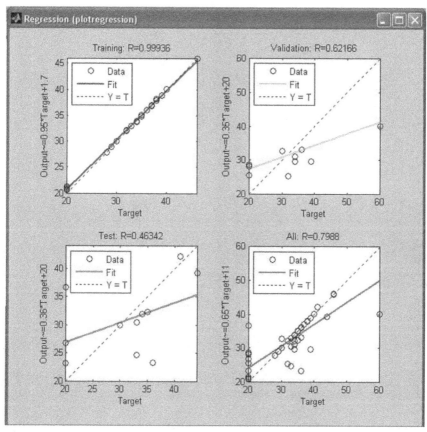

7th training set

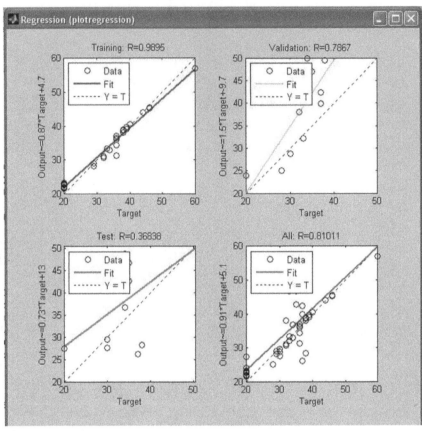

8th training set

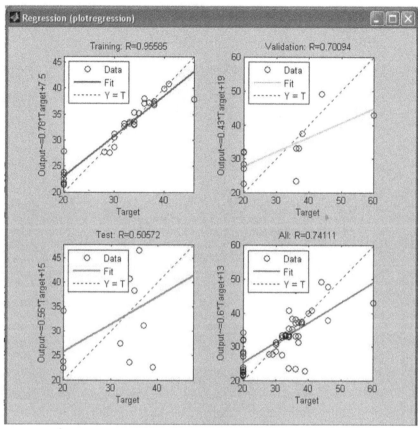

9th training set

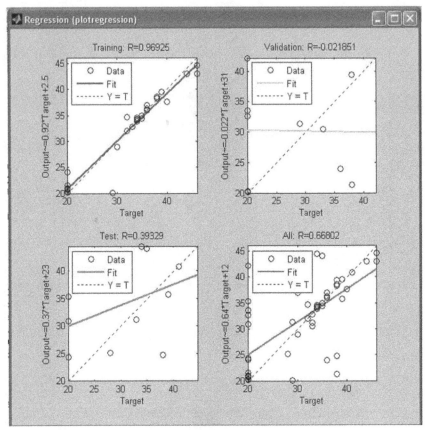

10th training set

Appendix D.2. Detailed statistical results of the 10 fold training set incorporating training subset, validation subset and testing subset excluding temperature input.

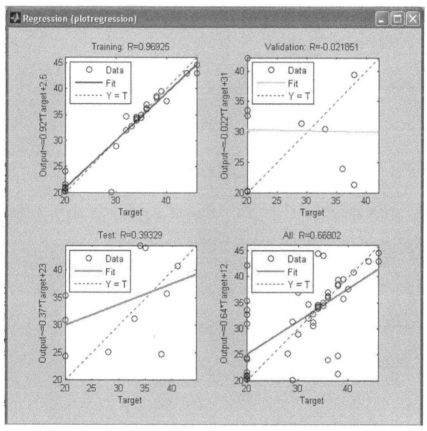

1st training set

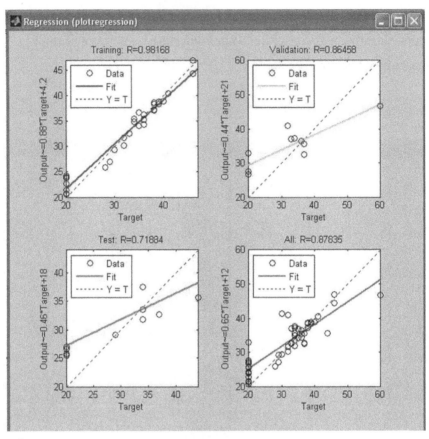

2nd training set

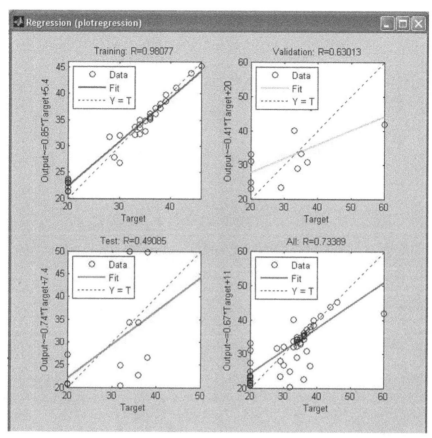

3rd training set

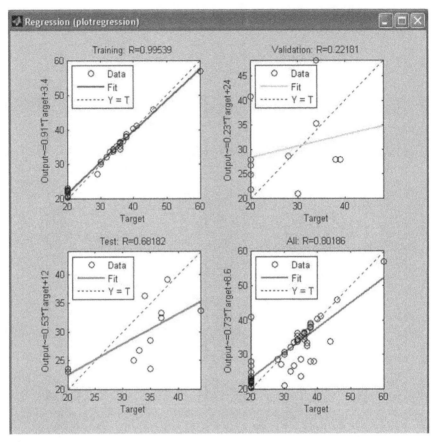

4th training set

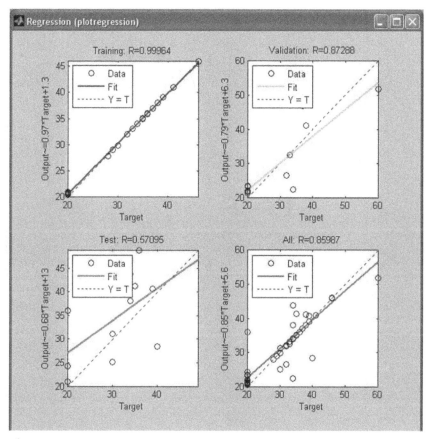

5th training set

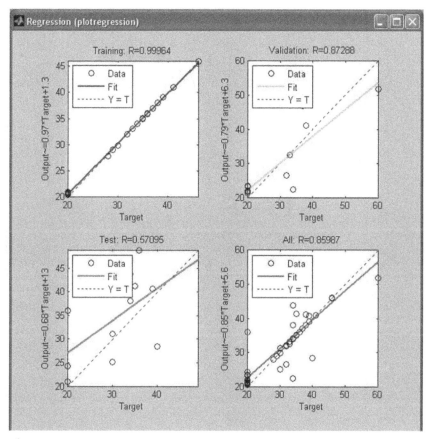

5th training set

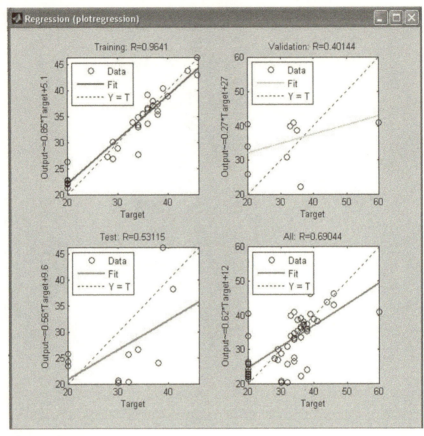

6th training set

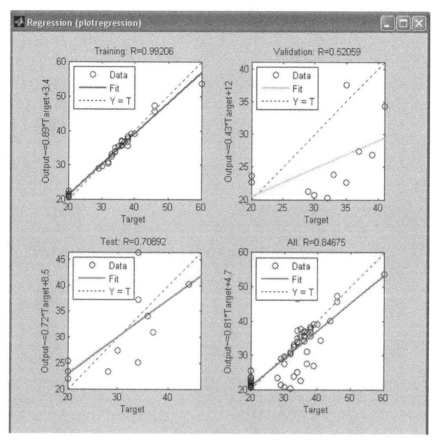

7th training set

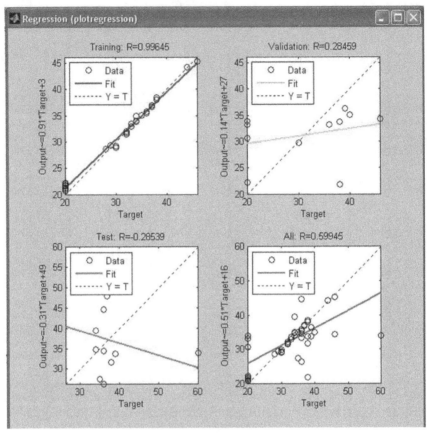

8th training set

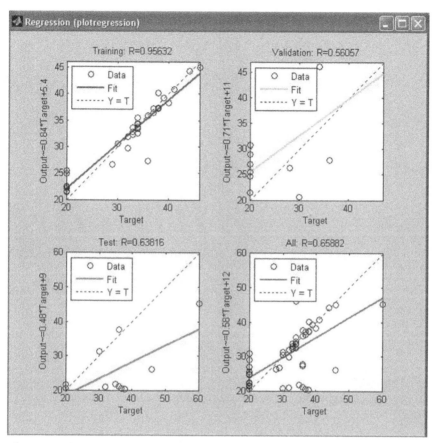

9th training set

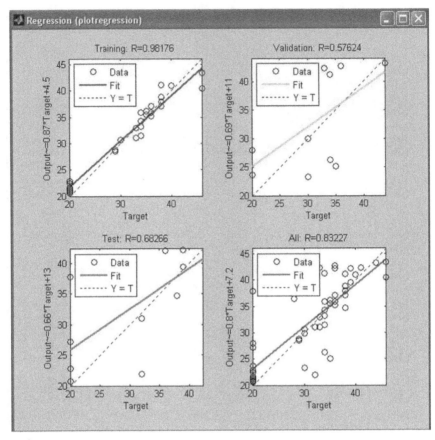

10th training set